禅意东方 X

居住空间

Feel the Eastern Zen Style — Living Space

欧朋文化 策划
黄滢 马勇 主编

华中科技大学出版社
http://www.hustp.com
中国·武汉

前言 Foreword

传统文化融入现代设计

达观设计：凌子达、杨家瑀

达观，一家来自台湾的优秀设计公司，两位主创设计师凌子达与杨家瑀，以融合东方文化特质，造型别具一格的现代设计，将事业推向成功的高峰，不但作品有口皆碑，同时在国际舞台上也是屡获大奖，将中国文化以设计为载体推向国际舞台，赢得业界的赞誉和尊重。达观也是《禅意东方》的老朋友，几乎每一期都能见到达观的作品，他们的设计以其独有的姿态层层绽放，令人回味无穷。很高兴在本书推出值得纪念的第十期时，与大家分享达观的设计观。

2014 美国环球杰出大奖获 "3 个类别最佳"
2014 德国红点奖获 "红点奖"
2014 日本 JCD 设计大奖获 "最佳 100"
2013-2014 意大利 A' 设计大奖获 "白金奖"
2013 年德国 iF 设计大奖获 "传达设计大奖"
2013 英国 FX 国际室内设计大奖获 "年度最佳优胜大奖"
2013 英国 SBID 国际设计大奖获 "最佳设计奖"

中国传统文化讲究道法自然，追求天人合一的人与物的关系。若从建筑设计的角度来品味中国传统文化，你就会看到一种平衡、和谐的关系。比如苏州园林，前人的建筑、园林和室内，充分利用光线、植物等自然条件，达到物同人在，人与自然物和谐相处的境界。其实这就是一种平衡，平衡其实就是一种协调，协调之后才会平衡。我一直以传统文化为根基，力求在空间中实现人与物的平衡，希望能够将建筑、室内、自然景观三方面融合在一起，成为互相观照、互为支持的统一体。

在我看来，中国传统文化中的很多精髓都可以被融入进现代设计中，使现代建筑更具文化底蕴，让传统文化绽放出时尚光彩。例如我司近日获红点奖的 "万科麓城" 售楼部设计项目，就是将 "太极" 中的 "阴" "阳" 元素运用到建筑的立面上。传统图腾与现代建筑相碰撞，相辅相生，毫无违和感。还有 "十方别墅" 的空间设计，也是导入中国传统文化思想，将 "风水" 与 "水墨画" 融入空间表现当中，使得风、光、水在室内形成良好的微环境，室内外景观互相映衬，还有水墨画彰显出空间的宁静、雅致、令人回味悠长。

空间设计要解决的核心问题是居住在中间的人是否舒适，室内设计最重要的是平面规划。一个好的平面规划，必须考虑功能问题，规划出合理的布局、动线，让在其间生活的人，感到舒服愉悦。我们在做设计的时候，很多时候都在弥补建筑设计的缺陷，优化户型的设计，希望能够让建筑、室内与自然景观融为一体。有了好的平面规划之后，至于是选择做中式，还是选择做欧式，那只是客户的审美偏好而已。

许多国际设计师的作品都是在表现自己的文化和生活方式。我希望能够在国际舞台中呈现出东方文化的作品，让大众可以从作品中看到属于中国人自己的东西，自己的风格。我会比较喜欢以一些新的建筑手法来呈现，建筑专业的背景让自己在设计时会从三度空间的角度去考虑问题，对于室内和室外环境之间的沟通和对话也更为敏感，能够更好地把握整体方向。建筑设计很注重建筑与周边环境的关系，这种理念也与中国文化中的 "风水" 相契合，我在设计的时候会把这种手法运用在室内设计中，关注与周围环境的互动。而做室内设计可以从建筑空间的角度去看问题，如设计一个墙面时需考虑它与周边的延伸，把一个立面当成一个平面去设计，所以设计作品会较为立体干净，有着很强的建筑感。说起风格，虽然做过不少奢华的欧式风格设计，但自己更倾向于简约主义。我们很喜欢做一些新中式风格的建筑与空间。从近几年参与的国际奖项可以看出，西方设计界已经开始注意到东方设计文化的兴起。华人设计师在国际的影响力越来越大。不管怎么说，中国人要努力，可以学习别人的长处，但是自身的成长、不断超越才是最重要的。

"万科麓城" 售楼部

返本归真

最近看了湖南卫视《天天向上》谭盾的专场节目，平添无限感慨。他对音乐的领悟力令人羡慕，他以自然为乐器，在天地之间，在传统文化中找到音乐的本体，将之融化到自身的音乐创作中，最终形成了个性化的音乐语言。他以水为音，以纸为音，以陶为音，在大自然的元素中寻找声音的起伏与韵味，自然是如此迷人，各有一具独特的灵魂，即使是同样的陶土，不同的质地，如红土、白土、黄土做出来的器具，发出的声音也各有不同，正合了一句古话"万物有灵"，对谭盾来说是"万物有音"吧，难怪汪涵说他不是音乐大师，而是音乐巫师。当他说到李白诗句"大音自成曲，但奏无弦琴"一句时，忽然击中了我的心。大自然孕育万物，经过亿万年时间的打磨，哪一件都是符合自然的，不但各种生物是有意识与个性的，连自然界的山川、流水、浮云、石头、土地也都有自己的型格，如果我们能从流水、虫语、花开、鸟鸣，听懂大自然演奏的无弦音乐，从而领略天地之道，那么也能从日升月落、山回水绕、四季更替中体会自然规律，从而一窥天人感应之境。天地自然，原本是我们最好的老师，连我们人的本身也是从自然中来，最终也要回到自然中去。既然自然是我们生命的起点和终点，那么在我们生命的过程中，所有的体验如果能符合自然的特质，才是最舒适最健康最美妙的。

说到音乐，不由得联想起喜多郎与班得瑞。喜多郎是一位永远不能被忽视的音乐鬼才。他从未受过音乐方面的教育，甚至看不懂五线谱，但他擅用电子合成音乐，将古典、流行、爵士融入电子音乐，他说："我可以用合成器创造海洋、冬天的海岸、夏天的海滩上的全部景色。"喜多郎的曲子韵律自成一格，他对东方文化有一种近乎本能的认知和理解，音乐中带着宗教般的神圣。他的音乐旋律优美，意境深邃，你可以从他的音乐中轻而易举地体悟到自然的雄浑、荒远，人情的温暖、冷漠，思维的理性、疏离、感动等极其抽象的意识形态。也可以通过他的音乐触摸到自然、历史、文化等这些用语言也难以说清的内容。从《丝绸之路》《敦煌》《故乡的原风景》到《宋家皇朝》每一部都在述说着一个故事，或一段往事，让人深深沉醉其中。此外，他对自然的领悟也是别具一格的，比如《大蛇》《凝露之舞》《川流之中》《风的声音》《水的精灵》《天地创造神》，那种神秘、深远、凝聚着雄厚力量的音乐让人感慨万千。

班得瑞，被称为"来自瑞士一尘不染的音符"，是由一群爱好生命的作曲家、演奏家、音源采样工程师等青年才俊组成。作品以环境音乐为主，亦有一些改编自欧美乡村音乐的乐曲，另外还有相当数量的作品是重新演绎的成名曲目。班得瑞最独特之处在于每当执行音乐制作时，从头到尾都深居在阿尔卑斯山林中，坚持不掺杂一丝毫的人工混音，直到母带完成。置身于欧洲山野中，让班得瑞拥有源源不绝的创作灵感，也找寻到自然脱俗的音质。虫鸣、鸟语、落花、流水，都是深入山林、湖泊，走访瑞士的阿尔卑斯山、罗春湖畔等处实地纪录，因而有感人至深的纯净与灵动。聆听《仙境》作品，让人以最轻松的姿态，仿佛进入梦幻般的天地之中，可以在柔软的草地上打滚，那里天地是透明的，风是香的，云是甜的，山是柔的，水是笑的，花儿是林中仙子，鱼虫是自然精灵，在天地之间与你嬉戏。

其实一个个音符本身是没有情感的，但是通过音乐人的精心编排和灵性诠释，融汇不同乐器的特质，通过高低起伏、快慢疏密的组合，形成独特的节奏和音韵，最终成为拨动心弦的乐曲。为什么有的音乐能够感动人，甚至能左右人的喜怒哀乐，而有的音乐却像噪音侵袭，让人无法忍耐而避之不及呢。我想可能是音乐本身无形无色，它通过耳朵直达我们的心灵，引发我们最纯粹的体验与感悟，因为剥去一切的包装与外界的干扰，我们只凭心灵去体验，当它与我们的情感、心情、性格相合拍，就会引发共鸣，甚至影响我们的感观、情绪和思维。音乐的本质，与禅宗所说的"直指人心，见性成佛"实在是很相似。

从音乐回到设计，大道至简，又殊途同归。有人说室内设计可以分成三等：初等的是实现功能用途，就是能用；中等的是有鲜明的风格，就是能看；上等的是体现文化品位，就是能品。我看在这之上还应该有一层，就是最好的设计能体现天地之道，自然规律，生命属性，让人心境平和，轻松愉悦，就是能定。

室内设计的发展和进步，和人的思想成熟进程是一致的。先是要解决实际功用，当功能不断完善时，就同步重视起美来；一开始对美的追求是繁复而悦目的，希望被重视被仰慕，接下来学会内敛与品质，可以说是低调的内敛和含蓄的张扬；当人生阅历到达一定高度后，对美的追求不再浮于表面，开始注重文化积蓄，品位修养和思想的传承，人生到此已值得尊重；如果能突破思想的局限，更上一步，不再关注世人的眼光与看法，真正懂得什么是自己想要的，什么是可以放下和舍弃的，什么是值得珍重和护持的，这种已经是近乎道。就如青原行思提出参禅的三重境界如出一辙：参禅之初，看山是山，看水是水；禅有悟时，看山不是山，看水不是水；禅中彻悟，看山仍是山，看水仍是水。当你看懂事物的本源和本体时，外表的包装与美饰已经不是那么重要，所需的只是去芜存菁与贴切适意。

那室内设计如何实现更高层次的追求。我觉得设计不能一概而论，人生就如登山，到什么山唱什么歌，走到山的哪一层，就看哪一层的风景，站在山脚的人，还无法描绘出山高人为峰的顶层风光。不同的人有不同的性格、喜好与观感，即使同一个人在不同的环境、境遇与心境下，他的需求、审美与思想都是不同的，所以好的设计师要做的不是改变客户的需求，而是在完善功能用途的基础上，适度提升客户的眼界，帮助他看到更好的生活形态，为将来预留成长的空间，但是不必拔苗助长，过犹不及。要美化给他美化，要风格给他风格，不同的风格混搭也完全没有问题，只要能做到和谐共处、健康生态、格调优雅。

其实真正做到道法自然的空间又何其不易，设计者不但要懂得设计中的种种原理、比例、用材、审美、新科技、新技术，真正好的设计要以环境为基础，以人为本展开设计，考虑风、光、景、绿、空气、磁场、微循环等种种要素，甚至是客户的喜好、健康状态、家庭关系、心理需求等等抽象要素，也要纳入整体的设计思维中，设计者要懂得物理学、风水学、心理学、养生学等等，如果在设计上能达到佛家的"八识"（眼、耳、鼻、舌、身、意、末那、阿赖耶）圆满，设计师已不仅仅是设计师，简直可以称之为"半仙"了。

在滚滚红尘中忙碌的人们，并不是每个人都能够懂得和欣赏返本归真时呈现出来的质朴、通透、简约、宁静的状态，不同的人有不同的需求和倾向，这个世界才能呈现出千姿百态，这样也挺好。返本归真，放下不必要的执念，欣赏和包容世界百态，在自然中品味生命的真谛与情趣，有空念念王维的《山居秋暝》吧：

空山新雨后，天气晚来秋。
明月松间照，清泉石上流。
竹喧归浣女，莲动下渔舟。
随意春芳歇，王孙自可留。

本书手绘花鸟图，来自清代余稚代表作。

目录 Contents

居住，返本归真
Dwell to Nature

008	国际化笔触设计，引领时尚品位生活	Global Design Leading Fashion and Taste
018	日照玉楼花似锦，楼上醉和春色寝	Sunlight on Jade Building, Flowers Are like Brocade; Intoxication Upstairs Echoes with Sleep in Spring Scenery
028	传统与现代相融，文化与艺术相生	Fusion of the Tradition and Modern, Coexistence of Culture and Art
040	墨染	Ink and Wash
048	界无限，观十方	Unlimited Boundary, Infinite Vision
054	玉蕴美德，谦若君子	Virtues of Jade, Be Like a Gentleman
066	无何有之乡，逍遥游之墅	Where Is Your Home? This Villa Is Right Here Waiting for You
074	山居：空灵的气息，演绎裸心的生活	Life in Mountains: Chilly Air Makes Heart Peace and Quiet
080	衍生	Generative
090	艺术地生活，诗意地栖居	Live Artistic, Dwell Poetic
102	师法自然，木石有情	To Follow the Nature, Wood and Stone Become Affectionate
112	隐在湖边，归在田园	Reclusive on Lakeshore, Returning to Countryside
120	执笔缱绻书古意，尽将流光付闲	Writing Is Attached to a Description of Ancient Conception; Time Can Be Spent at Leisure
128	以意带形，提纯人居美学	With Connotation to Promote Physical Shape, Dwelling Aesthetics Is Upgraded
138	浸染诗意的家园	Dyed in a Poetic Home
146	新传统文化视野下的精致文化生活韵味	Fine Culture and Charming Life by New Traditional Perspective
152	京华烟云	Moment in Peking
156	心闲不染尘，家藏笔墨香	A Heart at Leisure Is Free of Dust; Brush Pen and Ink at Home Are Fragrant
162	阅尽千帆，情系东方	View of Sails, Passion for East
166	水墨淡雅，回味悠长	Chinese Ink Painting Is Quite Refined and Worth Aftertastes
172	一曲古韵，千年风雅	Ancient Rhythm, Everlasting Grace
180	花将色不染，身与心俱闲	Inner Peace and Quiet
184	当时明月在，曾照彩云归	The Ever Bright Moon Escorts the Cloud to Return
190	谁解茶中味，水墨也生香	Who Appreciates Tea's Taste? Chinese Ink Painting Generates Fragrance
196	璧月琼枝梦秦淮	On Night When the Moon Waves Tree Branches, Qin-Huai River Is Dreamed
200	靛蓝与明黄铺陈的色彩华章	Graceful Poem by Indigo and Brilliant Yellow
204	悠游写意，水墨人生	Freehand Chinese Ink Painting
210	桃花源里藏芳华，美在深闺惹人醉	Springtime Is Hidden in the Peach Garden, Beauty Is Intoxicating in the Boudoir
216	冷色系打造，时尚花开	Cool Color Creates Fashion Prosperous

216	简约苏州印象，穿越文人意境	To Simplify Impression of Suzhou, to Traverse Prospect of Literati
226	鲜嫩欲滴的活力家居，令人怦然心动	Home Furnishing of Vigor and Vitality Makes People Eager to Accomplish Something with Excitement
230	风华内敛，气度雍容	Reserved and Graceful
234	亲近自然，轻松怡然	Get Close to Nature, Easy and Happy
238	岭南风情韵悠悠	Amorous Feelings Come Long-Drawn-Out From the South of Five Ridges
242	凭水临风逍遥居	Fancinating Residence
248	隐于山林，归于自然	Reclusive amid Forest and Mountain, Returning to Nature
252	醉卧花间君莫笑	Don't Laugh at People, if They Lie Drunk in Flowers
260	水映明月满庭芳	The Moon Is Reflected in Water, the Fragrance Is Stuffing the Patio
264	蝶舞香飞水云间	Above Water and Over Cloud are Flying Butterfly and Fragrance

酒店，因地制宜 Hotel to Local Conditions

272	传统与现代融合，开启一场震撼心灵的视觉盛宴
282	融合南亚风情，书写浪漫故事
294	樱花绽放迎嘉宾
302	西子湖畔的繁华隐贵
310	热情西藏风情，跃动多姿文化艺术

272	Fusion of Tradition and Modern Starts a Heart-Quake Visual Feast
282	To Blend South Asian Style, to Make Romantic Story
294	Cherry blossom to Greet Guests
302	Reclusive Noble on the Bank of Xihu Lake
310	Tibet of Ardor Motivates Diversity of Culture and Art

居住，返本归真

Dwell to Nature

国际化笔触设计，引领时尚品位生活

Global Design Leading Fashion and Taste

项目名称：绿地·海域观园样板间
设计公司：上海风锐设计工程有限公司
设计师：胡斐
摄影师：三像摄建筑摄影机构 张静

Project Name: Greenland · Sea-Viewing Show Flat
Design Company: FM.P Associates
Designer: Hu Fei
Photographer: Threeimages Zhang Jing

该户型样板间为典型的六合院布局，围合共享天地景致共创邻里情怀的中央庭院，半私密的入户节点和半公共的邻里空间，唤醒东方住居的仪式感和人文意义。户内规划南院、北院、中院和下沉式庭院围合成一墅四小院的格局，联合附赠的超大露台，精造复合院落的五重景致，实现东方有天有地的居住梦想；内部通过连廊打通南北进深，更围合成采光天井或私家内院，让居者于户内亲近天地。从双入户原木大门进入，映入眼帘的是更为国际化的样板间风格，明亮的内装色调衬托近13.7米超大南向格局和视野，客厅与餐厅灵活错层组合，共享庭院以及园林风景；二层三间卧室全朝南，尽纳阳光与自然；附赠屋顶大花园、超大地下室、双地下独立车库，大格局足够三代同堂舒适居住。

空间装饰以前瞻性的笔触来设计时尚尖端生活，引领国际化风尚。客厅家具崇尚简约、时尚、大方，造型独特，色彩高贵端庄，配合几何图案的地毯，简单勾勒，轻而易举突显时尚风格，使空间变得生动而鲜明。马赛克天花板下四盏别致的方形吊灯最牵引人的目光，如田字的造型，向人们诠释着优美的艺术气质。华灯璀璨，于纯净浪漫中透射出妖娆的光芒，一如有着迷人面孔般的"诱惑天使"，将火树银花般的华美发挥到了极致，流露出高贵的生活品位。为避免空间过于单调，设计同时穿插绿色植物、中式窗格、雕塑摆设等装饰元素，力求呈现独特的韵味。

餐厅与客厅灵活错层组合，以黑色与灰色等冷色调为主，简约而不张扬，于低调中彰显品质。餐厅中还设置有日本寿司料理台，利用花艺、枯枝等营造出温馨的文化氛围。餐厅装饰不追求繁复，采用或黑白或低彩度的饰品来点缀空间，

如墙面的水墨挂画、瓷器等,简约而不失时尚。

设计师还特别规划了一间时装设计工作室,布置自由灵活。一张原木大桌子坐立中央,桌子的浅褐色与原木质感使空间显得更自由、清新。空间以白色为主,桌子上、靠墙陈列了许多当季的包包、首饰,耀眼迷人,就像是一个专业的showroom。当然,在这里,那些女人们的最爱——包包、首饰,才是主角。

主卧宽敞、明朗、大气,洗浴室与休息区利用透明玻璃间隔开来,让两个不同的功能区都能拥有相对独立的空间,同时又保持空间上的一种通透性。床是休息区的主角,是一个让人卸下疲惫、放松身心的温馨港湾,因为人的一生当中,有三分之一的时间是在床上度过的。因此,设计师选择了一铺带着浓浓东南亚风情的床具,舒适、自然、保证了优质的睡眠质量。高高的床柱体现了一种低调的奢华,床架上的帷幔自由飘摆,更有一种来自热带风情的韵味,为空间增添了不少的浪漫气氛。次卧纯净、舒适,黄色系的点缀,更显空间视觉之美,犹如花朵一样,在阳光灿烂的季节里温柔绽放。

A project this show flat that is divided into a 6-patio structure. The whole space surrounded with landscape and centered on the courtyard starts its prelude with a semi-private entrance and a semi-open neighboring community. All are intended for rousing the ritual dwelling in the east and the cultural connation. Patios in direction of south, north and middle and a sinking one embrace a villa. Additionally, a super large balcony is offered for free. Corridor winds throughout from south to north, making a daylighting patio or a private inner courtyard, within which occupants can get affinity with the heaven and the land.

Through the double wooden door, comes the interior designed with global standard and bright hue to set off the super large pattern and view. The living room and the dining room are arranged on staggered floors, both facing the courtyard and the garden. The former's luminous brightness sets off its 13.7 meters width and south-oriented view. 3 bedrooms on the second floor face south to take in abundant sunlight. The whole space with the large roof garden, the basement and the double garages underground make the families of three generations live under the same roof.

Spatial decoration accomplishes fashion and top life in a leading and pioneering approach. Furnishings in the living room are simple, fashion and grand with their unique modeling and noble colors to match the geometric carpet. All are easy and convenient to sketch out and set off the fashion style, the space thus becoming vivid and bright. Four square droplights down the mosaic ceiling are so eye-catching, like a large square consisting of four smaller ones to interpret good artistic sense. Splendid light glisters in peace and romantics, making a charming image of alluring angel. Beauty of flaming trees and silver flowers is exerted to the utmost, leading life taste to its summit. The implantation of greenery, window of lattice and sculpture accessories ground off the stiffness of design to allow for flavor of another kind.

The staggered space of the dining room and the living room is coated in cold color, simple and reserved to highlight its quality. Japanese sushi workshop in the dining room makes the best use of flowers and dried branches to accomplish a warm cultural atmosphere. Decoration in the living room is not aimed for complication, but embellished with black-white or low saturation to look both simple yet fashion, like wall painting of Chinese ink and porcelain.

There is also a fashion studio, flexible and full sets of functions. In the center is a log table, whose light brown and wood texture make the space much freer and more refreshing. The space is dominate in white. On the table and along the wall are seasonal bags and jewels. Here therefore looks like a professional showroom. Of courses, bags and jewels, women's favorite serve as the leading actor.

The master bedroom is broad and bright. Between the bathing area and the resting area is transparent glass to allow for relatively independence and privacy. Bed, the protagonist of the resting area, is a refuge to relax body and rejuvenate soul and mediate mind. One third time is spent

on sleeping on bed. The bed here is of Southeastern style, comfortable and natural to ensure a good sleep. High bed posters embody a conservative luxury. Drapers on the bed shelf swing now and then, bringing in tropical feelings in increasing spatial romantics. The secondary bedroom is pure, equally comfortable, whose yellow hue compliments the visual effect. The room thus feels like a flower, blossoming in sunny time.

Feel the Eastern Zen Style — Living Space X

日照玉楼花似锦，
楼上醉和春色寝

Sunlight on Jade Building,
Flowers Are like Brocade;
Intoxication Upstairs Echoes
with Sleep in Spring Scenery

项目名称：深圳招商鲸山觐海九期流水别墅示范单位
设计公司：深圳市朗联设计顾问有限公司
设计师：秦岳明
摄影师：井旭峰

Project Name: Water Flowing Villa
Design Company: Shenzhen Rongor Design and Consultant Co., Ltd.
Designer: Qin Yueming
Photographer: Jing Xufeng

东汉许慎《说文解字》说："玉、石之美者。有五德：润泽以温，仁之方也；鳃理自外，可以知中，义之方也；其声舒扬，专以远闻，智之方也；不挠而折，勇之方也；锐廉而不技，洁之方也。"在本套招商"鲸山觐海"九期流水别墅中，设计师以"玉"为主题贯穿空间，运用传统开合、迂回的技艺手法展现空间的韵律与趣味。通过节奏的变化与跳跃来体现不同空间的韵律之美，赋予空间一种别致的情调。

在本案新中式风格中，时尚家居环境将怀旧的情愫融入其中，无论是清晨朝阳，还是黄昏暮光，都让人沉醉在过往岁月的奢华中，平静似水。即便经过光阴的洗礼，依然安之若素。徜徉其中，浅浅斟酌，写满流年，低吟浅唱。原来，生活可以是一种心情。

居室蓄满缱绻诗意的中式纹样，清雅隽秀的中式家具，宁静的瓷器、玉器，精美绝伦的官扇，中国水墨画的宫灯，素笺上画笔勾勒的墨痕，画下最痴情、写意的一笔，潜藏在内心深处那属于古典优雅的诗情，只有热爱中式文化的人才能懂得，如那工笔画中的花香，摇曳着美丽和优雅，踏过尘埃的声音，携一袭温馨的梦想，无边的风月，写满流年里最风韵的念想。

幽幽的灯光，透过印花丝绸的经纬，让新中式居室承载了熏香的思念。凝眉处，中式花格将往日流光过滤；漏窗处，一盏水墨画古灯闪着幽幽的魅光。原来，有一种意境，叫做"无声胜有声"。

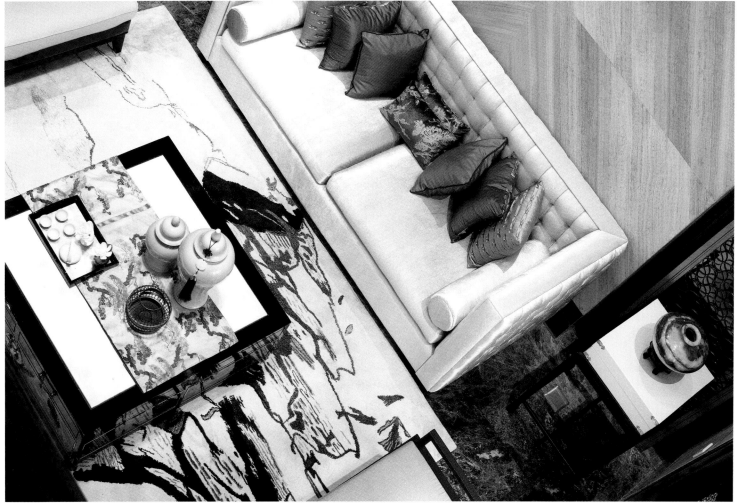

The book of *Word and Expression* by Xushen in East Hang Dynasty describes that, jade, a beautiful stone has five virtues. Namely they are benevolence, righteousness, wisdom, courage and purity. And throughout this project is carried the theme of jade. With traditional approaches, open and close, and roundabout, spatial rhyme and fun are expressed. The changing and jumping rhyme is intended for different spatial aesthetics, allowing for a unique sentiment.

Into the modern setting are nostalgic feelings, which whether at dawn or at dusk, makes people lost in past luxury but with a peaceful state of mind. Despite time passing, people can still be at ease and in leisure. It's here that life can get into peace and quiet. Chinese pattern is deeply attached to the living space. Chinese furnishings and accessories are like ink trace on white paper, like porcelain, jade article, fan widely popular in ancient officers, and palace lantern. The classical and graceful poem hidden in the depth of heart can only be perceived by people loving Chinese culture. Flowers fragrant in the elaborate-style painting offer peace and quiet that has been through years but still appealing and attractive.

Light goes faint through printed silk, and the neo-Chinese interior becomes a carrier of incense. Naturally, sights are fixed onto window of lattice that seems to have tided the past memory. By ornamental perforated window is a painting of Chinese ink with dim light. That accomplishes a mood of silence being better than voice.

传统与现代相融，文化与艺术相生

Fusion of the Tradition and Modern, Coexistence of Culture and Art

项目名称：中海苏州独墅湖别墅
设计公司：HSD 水平线室内设计
设计师：琚宾
参与设计：韦金晶、唐静静、夏晖、吴圳华、张旭
摄影师：孙翔宇
撰文：Sam
面积：800 m²

Project Name: CSCL (Suzhou) Dushu Lake Villa
Design Company: HSD Horizontal Interior Design
Designer: Ju Bin
Participant: Wei Jinjing, Tang Jingjing, Xia Hui, Wu Zhenhua, Zhang Xu
Photographer: Sun Xiangyu
Text: Sam
Area: 800 m²

艺术与设计一直是装修时人们所关注的重点对象，苏州独墅湖别墅，坐落于江南地域特征浓郁的建筑群里，设计希望在现代人居的习惯与浓郁的地域文化之间寻求平衡，希望通过对空间的设计与整合表达对自然的礼赞。

如果一个空间蕴藏着一种难以言明的空间气息，或现代，或异域，远离繁华，回归自然，里外一致，置身于静谧，又何不是另一种享受。针对该项目，设计师试着让异域和现代文化在碰撞中相融汇，通过深厚的艺术底蕴，开放、创新的设计思想与尊贵的姿容，向世人表达从简到繁，从整体到局部的精雕细琢。从材料到色彩，从家具到装饰品，在本案中我们都能感觉到强烈的历史感与浑

厚的文化底蕴。

如果说建筑是设计的实质，那么室内空间则是精神的表现，设计师在本项目中试图延续的不仅仅是建筑的表象，而是建筑的精神。在分析光影、水路和体块与建筑联系与分化的关系后，以抽象的方式引导着宽敞开阔的视野；对节点的细致处理也体现出细腻的设计，天然材料的使用既让所有节点得以最好的显现，又巧妙地体现了其工艺价值。所有设计思想、设计风格，都表达着设计师对生活的态度。这里没有壁炉、水晶宫灯，也没有罗马古柱，但处处都透露着宜居的氛围，充满着古典主义的精髓与异域的琢磨不定。在这个功能性强并且风景优美的建筑里，设计师敏锐地把握客户的需求，在不浮夸的前提下设计了一套大方得体的优雅居所。

无论是家具还是配饰均以其优雅、唯美的姿态，平和而富有内涵的气韵，描绘出居室主人高雅、贵族的身份。室内充满着白色、金色、黄色、暗红与黑色，显露出高雅与和谐，当我们摒弃了江南的白墙黑瓦，随之而来的则是明亮色彩。开放宽容的设计态度，让人们在这里生活丝毫不显局促。进入客厅，一大张色彩斑斓的地毯吸引着人们的眼球，不规则的几何织纹、跳跃的拼接色，异域风情也不外乎如此。木料与石料拼接的圆茶几、背墙摆放的巨型低矮沙发、米灰色绵软的坐垫、色彩斑斓图案奇异的靠垫、木材与布衣组合的躺椅……慵懒、惬意、精致、美艳。

Art and Design have been paid close attention to in time of decoration. Dushu Lake Villa, a compound located in Suzhou of low reaches of Yangtze River and abundant in China southern geographical features, is aimed to make a balance between modern living habit and rich local culture, and pay homage to nature by integrating design and space.

If it bears an implicit space atmosphere that's beyond description, modern, exotic, or far away from downtown to allow for peace and quiet, a space is bound to be a great enjoyment. A project this space makes that is a trial exploration to unify exotic and modern culture in their collision. The rich concept, the open and innovation design philosophy and the noble appearance vary from the simple to the complicated, from the whole to the part, from the material to the hue, and from furniture to the accessory. All exert a strong sense of history and tradition and a rich cultural heritage

for people to feel.

If the building is the essence of design, then the interior is the manifestation of spirit. The continuation in this project is not only the appearance of building, but its spirit. The analysis of the relationship between light and shadow, and that between waterways and building blocks, leads to an abstract way to guide the spacious to be open and horizons; the treatment of nodes reflect the careful handling of delicate design, using natural materials to guarantee that nodes can be shown best, so that technology value can be given play to ingenious. All design ideas and styles express designer's attitude towards life. Without fireplace, or crystal palace lights, or ancient Roman columns, everywhere reveals a livable setting for modern people, which is filled with classical essence and something exotic and elusive. In such a highly functional and scenery building, a decent and elegant home is made based on the keen grasp of customer needs, all without exaggeration.

Whether furniture or accessories are elegant, beautiful, calm and full of meaning artistically, depicting that the owner's elegant, aristocratic identity. The interior filled with white, gold, yellow, dark red and black, confides in elegance and harmony everywhere. When southern white walls and black tiles have been abandoned, the bright colors are preferred. The Open and tolerant attitude makes a space that looks no longer significantly cramped. The living room features a large sheet of colorful flowery carpet, whose irregular geometric pattern and jumping colors feel nothing but exotic. The round mosaic coffee table of wood and stone, the giant long couch against the wall, the griege and the colorful-patterned cushions, and the deck chair of wood and fabric are bound to make people feel lazy and comfortable, for they are sophisticated and glamorous.

墨染
Ink and Wash

项目名称：陈宅
设计师：唐忠汉
摄影师：游宏祥
用材：橡木地板、铁件、石材
面积：200 m²

Project Name: Chen Mansion
Designer: Tang Zhonghan
Photographer: Kyle Yu
Materials: Oak Flooring, Ironware, Stone
Area: 200 m²

水墨，即墨，即无色，即一切一切。由于建筑本身的格局受基地格局动线的影响，故空间以一道长廊作为中心，渲染扩散至每个空间。位于廊道底端的书房空间，在设计规划之初将其刻意打开，让光线、空气得以流通。

空间在色调上企图采用介于深浅之间的中调色彩，如墨般晕染空间量体、柔和空间氛围，部分的铁件色调整合空间焦点、材质堆栈，亦如墨色的推演。

穿越廊道，格栅层层竖立，透过立面肌理，将视线引入各个主要空间。廊道延伸出轴向，强调出在端景赋予的空间窗景。公共领域以客、餐厅及中岛为同一轴线，开放的空间、明亮的厅堂，串联使用。私人领域使用平质素材，与景致融合，透过木质格栅过滤直接语汇，看见透光的优雅书房，石木的自然纹理渲染出空间的沉静安逸。

平面图 / Site Plan

The layout of the architecture due to the arrangement of movement by the site has fixed a long and connection corridor in the space as the center, exaggerating and proliferating itself to each space. The library at the end of the corridor, is deliberately made to open up at the beginning of design and planning, so that lighting and air can be proliferated and spread out from the spot.

Exaggeration as ink-color, of coloring hues, neutral coloring between lightness and darkness is tempted to be used, similar to body weight of space as if dyed with ink, forming tender space atmosphere. While coloring of partial iron pieces is adjusted to integrate spatial focus through overlapping of materials, showing ink-color derivation

With established layering grids along the corridor way, and through standing texture, they have introduced ones into each of the main spaces. As corridor way extends into axial direction, the vision should emphasize end-view that awards window-view for space, the public domains integrate living, dining, and kitchen island onto the same axis, with open space and bright living room, helping to link up various functions. As for private domains, flat and even materials are used, harmonized with the scenery. Direct glossaries are being filtered through wooden grid, making one to perceive the elegance of objects towards lighting. For natural texture of stone and wood in the library, air of humanity plays up the sense of serenity and quietude; through the drilling of light-beams, they have then spread out upon the wall-surfacing.

界无限，观十方
Unlimited Boundary, Infinite Vision

项目名称：十方别墅
设计公司：上海达观建筑工程事务所
设计师：凌子达、杨家瑀
面积：250 m²

Project Name: Shi Fang Villa
Design Company: Kris Lin Interior Design
Designer: Kris Lin, Yang Jiayu
Area: 250 m²

这是一个提供互动、等候、驻足、沉思、冥想、观赏和穿越的公共空间，旨在实现人与人、人与建筑、人与自然之间的交流与融合，所以企图把室内、建筑、景观三位整合成一体，完全的整合是本案设计的核心价值。

同时这也是一个建筑改造的项目，把原有建筑的室内隔间和建筑立面全部拆除，只保留结构体，全部重新来过，更提供了一个打破室内、建筑、景观分界线的契机。设计采用无框式的玻璃窗，让景观水池延伸到室内，室内的石材墙面延伸至建筑外墙。桥从室内穿过建筑外墙与水池，直达景观草地，各元素不断地交错融合，打破了空间界限。

设计以室内为核心，希望每个空间或走道能与自然结合，所以设计的墙体纯粹而且各自独立，同时透过无框玻璃呈现出大量的景观面，让室内与景观结合，站在室内的每一个位置都可看到优美的自然风景。

A public space this project makes where for people to interact, wait, ponder, meditate, reflect, appreciate and traverse. It's aimed for communication and blend for people and people, people and construction, and people and nature. So the interior, the construction and the landscape are transformed into one, where an all-around combination is the core value of this project.

Also a refurbished project this space is, whose facade and compartments have been demolished with the original structure kept. This provides a moment to break the stereotyped boundary among the interior, the construction and the landscape.

Through the frameless glass window, waterscape of pool has extended within, while internal stone wall finds its way onto the external wall. Through the wall, a bridge comes out across the pool until it reaches the lawn. With the constant blend, the fixed boundary is thus broken.

The interior taken as the center, all sections or each aisle is intended to be fused into the nature. So walls are designed independent or exactly on their own. The large-size frameless window glass allows for many landscape-enjoying positions, so the interior and the landscape are united, and you bet that wherever you are, you can have a good view of the nature.

玉蕴美德，谦若君子
Virtues of Jade, Be Like a Gentleman

项目名称：燕西华府
设计公司：HSD 水平线室内设计
设计师：琚宾
摄影师：井旭峰
面积：846 m²

Project Name: Wonderland Mansion
Design Company: HSD Horizontal Interior Design
Designer: Ju Bin
Photographer: Jing Xufeng
Area: 846 m²

本案以"玉蕴"为设计理念，将璞玉的气质、故宫传统的经典建筑形式，与度假的自然感觉相结合，透过玉石、木纹、金属、壁布等材料的契合，彰显当代东方美学气质。

设计师意欲将当代与东方，时尚与经典，内蕴与大气，共融为独特的东方美学气质，从线条到材质，从色彩到空间布局，将精致的细节与品质融入到空间中，展现出一种精致东方精神。

在空间的表现形式上，则从玉的五种自然属性来入手，将玉的质地、光泽、色彩、组织以及意蕴与空间的形式、材质、色调、景观一一对应。

其一，坚韧的质地，空间中强调竖线线条与空间体块微妙的层次之美；其二，晶润的光泽，应用漆面、玻璃、金属等材料，强调当代时尚与玉质的碰撞，呈现出符合当代审美情趣的空间；其三，灵动的色彩，空间中软装方面以优雅精致的面料和丰富的材质交相辉映，呈现空间的时尚与雅致；其四，致密而透明的组织，将传统建筑窗棂的形式重新解构，形成半透与不透的层次关系；其五，舒畅致远的声音，中庭自然的水体与光的倒影形成空间中的空间，似一曲"趣远之心"。户外景观的自然设计，移步换景的手法，给空间带来了丰富多变的视觉延伸，可观、可游、可赏，体现出度假式的自然情怀。

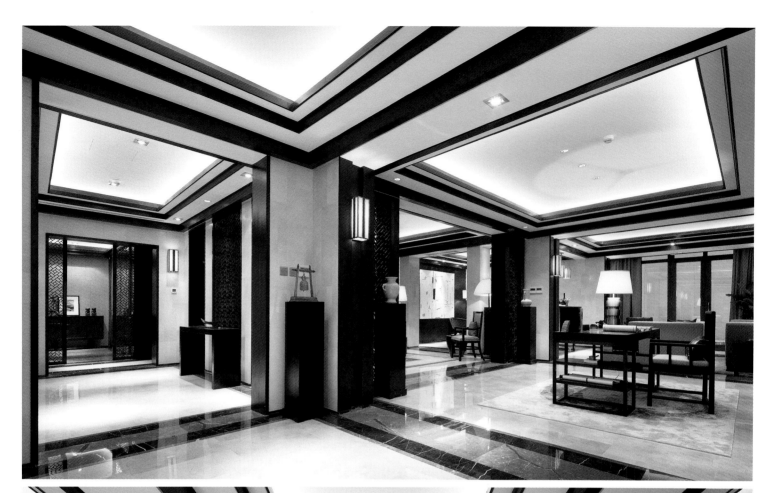

The space with "jade" as its concept merges jade nature and architecture form of the Forbidden City into the interior design, which based on holidays feelings convey the modern aesthetics of the east by using jade, wood, mental, and fabric.

According to the design, the modern and the east, the fashion and the classical, and the connotation and the magnificence are fused into the oriental aesthetics. All from color and layout, and from lines to materials, overspread details and quality into the space just for presenting a delicate spirit exclusive to the east.

Five prosperities of Jade are made full use of. Jade's quality, glister, color, issue and implication respectively echo with the form, the material, the hue and the landscape of the space.

Vertical liens and spatial blocks are stressed to reflect the toughness of jade, lacquered surfaces, glass and metallic materials to stress the collision of

contemporary fashion and jade to create a space that reconcile to today's aesthetics, fabric and materials set off with each other to symbolize Jade's color, the reinterpretation of window lattice that makes layers transparent and opaque to refer to Jade's issue and reflection of waterscape and light completing an image that there is another space yet to come out a space to provide a metaphor of sound. Design of landscape involves approaches that views change with steps goes. Visual effect is thus extended in the space where to watch, to appreciate and to enjoy in a resort setting.

无何有之乡，逍遥游之墅
Where Is Your Home? This Villa Is Right Here Waiting for You

项目名称：中海烟台紫御公馆别墅
设计师：琚宾
参与设计：黄智勇、唐静静、陈建业
用材：新西米、胡桃木、柚木饰面、黑钢
面积：431 m²

Project Name: Purple Royal Mansion House, Yantai
Designer: Ju Bin
Participant: Huang Zhiyong, Tang Jingjing, Chen Jianye
Materials: Marble, Walnut, Teak Veneer, Black Steel
Area: 431 m²

真实与虚空。"空间"本为哲学用语，是人类文明达到一定程度之后产生的概念。到了现代社会，随着人类不断的进化，文明高度发展，"空间"概念在不同的领域有着不同的定义。在紫御公馆项目里，"空间"被定义为建筑的内部。当人游走在建筑的内部空间中，其游走的领域正是真实与虚空之间，意为任何一个经过人的视线及行动范围中有阻隔和限制作用的实体，其为真实，名为"间"，其他的元素则为虚无，其为虚空，名为"空"。"空"与"间"的不同定义，在不同的使用功能中，体现出室内的建构。在项目中，除了楼层间的分隔以及满足特殊功能空间的要求外，极少引进"绝对分隔"的定义，大多采用的是"半分隔"的定义，使不同的功能区间中，视线、空气、光以及活动区域都能在"半分隔"的限定元素中隐约出现，给人以延伸之感。另外，本案大量采用了石材和木材，也在局部区间运用了金属材料，在"真实与虚空"的定义下，对这些材料加以拆分组合与运用，使之界定在东方传统和当代视觉相结合的形式中。理解和应用这些元素，正是设计师营造现代具有东方古典哲学体系的室内建构的核心。

在文化上，设计师希望此项目展现东方人谦逊和礼让的生活态度，在客户认可的条件下，设计师把这种谦逊和礼让的生活态度转化为空间上的"无"与"隐"。虚静、恬淡，但却充盈、灵动。本案以其空间结构进行诠释，不同的功能空间围绕着一个光盒子布局，自然光穿透在两层的中空空间上，弥补了建筑中心空

地下夹层平面图 / Basement Interlayer Plan

地下层平面图 / Basement Plan

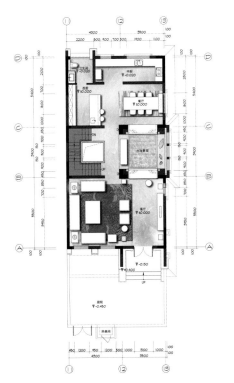

一层平面图 / First Floor Plan

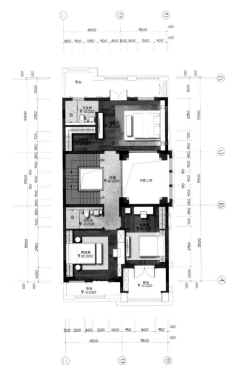

二层平面图 / Second Floor Plan

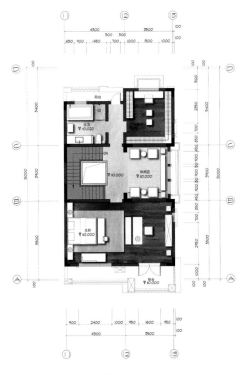

三层平面图 / Third Floor Plan

间难以采光的缺陷；同时在空间与空间的分隔上做了特殊处理，采用透与不透、隔而不透的分隔方式；在家具的选用上，主要采用了具有现代东方气质的简约家具和国际性的经典家具。

在物理空间意义上最理想的环境，其实就是一种人与空间的共处艺术——"游"。当人游走其中，其感悟的领域正是真实与虚空之间。空间中的实体及实体暗藏着的文化属性，具体物件与精神隐喻之间，将"虚""实"的概念结合到不同的使用功能中，对其作出不同的定义，并体现在室内的建构中，这种"游"的感悟或许能成为室内空间所寻觅的另一条美学出路。

Trueness and vanity are opposite for sure. The word of "space" was used to be in philosophy, a concept that developed as human culture accumulated. When human being step into the modern society, with the constant evolution and culture highly developed, space begins to get more connotations in different areas. And in this project, space's concept is fixed into the interior that changes from the true to the fake, where anything able to be caught into eyesight or block human steps is called partition pronouncing "Jian" in Chinese language, while other elements are called vanity sounding "Kong" in Chinese. Both together mean space. Such understanding to the two concepts combined into different functions to make various definitions. And they are then embodied into interior design. Semi-partition is frequently used instead of the absolute separation, except for the boundary among floors or particular functions to be met. So sight, air, light and activity can be looming among the semi-separate elements, visual effect thus extended. In addition, the extensive use of stone and wood, as well as the emergence of metallic materials somewhere are available between modern vision and east tradition. Perception and application of these elements, make the classical philosophical core of this project.

Culturally, this project is intended to make a trial exploration of humility and comity of the eastern society, which, approved by the customer, have been shifted into "nothing" and "seclusion".

A space here makes of emptiness and tranquil, and at the same time, it's flexible and smart. Here successfully makes a project whose definition is done by spatial structure, where different functional spaces are arranged around a light box, with daylight penetrating the two space to ground of shortness of light. Meanwhile, partition between is handled particularly to make the space adjacent transparent and opaque, and separable but still linked. As for furnishings and, the Oriental minimalist goes together with international classics.

The most physically ideal setting, in fact, is to express an art of coexistence between man and space. That is "tour", during whose course, people can get a good perception of trueness and the void and cultural property of concrete articles. Into specific objects of tea ware and spiritual metaphor, the concept true and void are combined, making different definitions, which are reflected into interior. Such treatment of touring sentiment may well accomplish another way for aesthetics realized in space.

山居：空灵的气息，演绎裸心的生活

Life in Mountains: Chilly Air Makes Heart Peace and Quiet

项目名称：滕州亿丰·和家园样板间 302 户型
设计公司：济南成象设计
软装公司：济南成象设计
设计师：岳蒙

Project Name: Yifeng·He Homelang Show Flat for 302 Unit, Tengzhou
Design Company: IMAC+NG Space Planning
Upholstering Company: IMAC+NG Space Planning
Designer: Yue Meng

"山居"是一种生活方式，远离尘世的喧嚣、安享隐居的静谧。细微的感动成就了生活的内容，多样的色彩协调着空间的存在。本户型以咖啡色为主色调，富有生命气息的绿色跳跃于空间的每个角落，有种"空山新雨后、天气晚来秋"的清爽与惬意。

推开门，首先映入眼帘的是客厅与走廊交界处造型独特富有张力的泰山石，它秉承了泰山的文化，延续着泰山的地域特色。跟随着探寻的脚步，视线不由自主地停留在了客厅中：现代中式的品牌家具、装饰画、水墨画地毯、贝壳壁纸、古铜色做旧的小鸟挂件、贝壳夹丝玻璃、不锈钢与亚克力结合的花槽内置绿色毛石头与花卉，富有自然气息的原木落地灯，现代感十足的方形水晶吊灯，日式铁壶，茶船，香炉与香灰……凡此种种，共同营造出一种悠然而静谧的真实生活感。

餐厅与客厅含情相望，餐桌上那热烈的黄色跳舞兰最为惹眼，热烈的黄使整个空间瞬间鲜活起来，更呼应了厨房的热情，把女主人在厨房为一家人忙碌的热情传递给整个空间。

书房充满了浓郁的书卷气，一方宽绰的书桌上笔、墨、纸、砚一应俱全，展开的画轴，极具文化气息的毛笔，古瓷底座台灯……无一不在述说着中式的情趣。书房背景墙被做成了整片的墙柜，用以收藏书籍、陈列艺术品，展现了书香的古典气息和艺术品的优雅品质。

主卧整体空间开阔，采光良好，阳光洒入，整个房间便洋溢在一片暖融融的氛围里。插上几株花卉，把自然的芬芳带入室内，静谧的美好带给人们一份意外的收获，与自然结合便有裸心的洒脱。卫生间的装饰画选用十二月份的场景画，将一年景致收藏于家中，展现出气象的变化之美存乎于生活的点滴之中。

儿童房为一喜欢航海的男孩房，整个房间围绕十几岁男孩的心理展开设计，包括挂在墙上的船舵模型，陈列于床头与书桌上的书籍、玩具、船舰模型、海贼王公仔等。从色彩心理学的角度考虑，空间整体色系为蓝色系，符合了十几岁男孩的色彩心理。

在这里，设计师表达了一种宁静的生活方式与自然的生活态度，让人们享受到一种超脱繁杂喧嚣的"山居"生活。

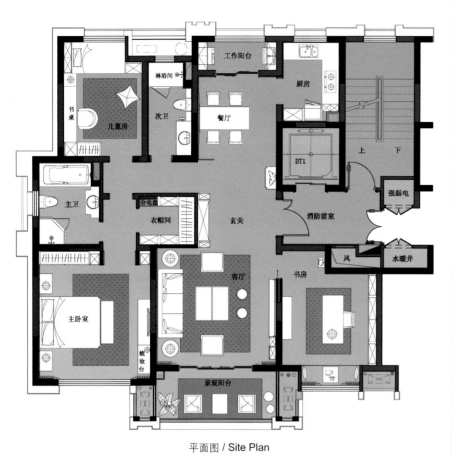

平面图 / Site Plan

"Mountain Dwelling" is actually a life style where to keep away from social hustling and busting. Life can be made by delicate touch. Spatial existences are stressed by various colors. A project this space is with coffee dominating the space and green jumping into everywhere, along which come a sense clear and bright, an image that rain has gone fresh and the hilly scenery is the most beautiful in autumn.

Between the living room and the corridor, the defining area features dramatic modeling of Taishan Stone to carry forward the local culture of Taishan Mountain, an element typical to be in accordance with Taishan Mt. culture. With steps to go forward, eyesight is natural to be focused on the living room, where brand furnishings are of modern Chinese style. Like decorative painting, carpet of wash inking, shell wallpaper, old-weathered brown bird pendant, shell of wire glass, log flooring lamp, square crystal droplight, Japanese iron teapot, censer and incense ashes, items you can name altogether make a real life sense of peace and quiet.

The dining room and the living room are echoing with each other, seeming to be looking at with loving eyes. The dancing lady orchid on the dining room makes the whole space come alive with its enthusiastic yellow in corresponding with the passion in the kitchen of the hostess occupying herself with cooking.

The study is certainly of scholar's style. On the large desk are brush pen, ink, paper and ink stone. The folded picture scroll, the cultural writing brush, and the table lamp with old china pedestal are all telling of their Chinese charming. The backdrop is designed into a whole wall cabinet for books and art pieces, catering for book fragrance and artistic grace. Broad and wide on the whole, the master bedroom takes in abundant daylight, which exposes the space into warm. Flowers and plants introduce natural aroma. The peace and quiet allows for surprises. Decorative paintings in the bathroom are themed on month scenes of 12 months in a year, a symbol that aesthetics by meteorological change exists in daily life.

The child's room is centered on the preference of the boy who has a sailing hobby, fully reflected by rudder model on the wall, books and toys relevant to sailing, vessel models and One Piece. The blue hue dominating the space is consistent with the color psychology of the teenage boy.

Here conveys a lifestyle of peace and quiet, making people enjoy dwelling in mountain, free of hustling and bustling and kept away from social life.

衍生
Generative

项目名称：台湾天母777蔡宅
设计公司：近境制作
设计师：唐忠汉
用材：石材、铁件、实木皮、镀钛、木地板、皮革
面积：383 m²

Project Name: Cai Mansion
Design Company: Da Design Apartment
Designer: Tang Zhonghan
Materials: Stone, Ironware, Solid Wood Veneer, Titanizing, Wood Flooring, Leather
Area: 383 m²

借由单纯的结构元素，重复、序列、扩大、转折、生成空间主要基地。
基地特性为一具有落地玻璃帷幕外墙的三面采光空间，在此空间优势下，利用垂直延展至天花的深色格栅作为调和空间的一种反向张力，在垂直律动的节奏中除了于表面反映日照，也利用立面进退处理的手法，使得墙体于不同视角中展现轻盈的线条感与厚实的体量感，大尺度的水平桌面，一方面界定了场域分属、金属材质的半反射质感、同时也让纵向格栅的比例在视觉上获得平衡。
在上层空间中，两道量体串联划分出空间的三种功能区块。在公共领域将量体脱离外墙玻璃帷幕，利用材料质地的延续，延伸空间视线，实现光影及空气的流动，在私人领域则利用镜面弱化墙面量体压迫感并反向延伸空间。

This is a project that has a main base of the space through repetition, sequence, enlargement, and transition of a single structural element.

The base is a three-sided lit space with full height glass panels. In this advantageous space, the dark grating is extended vertically to the ceiling in order to harmonize the reverse tension in the space. The vertical rhythm not only reflects the sunlight but also brings the line of the light and the solid volume effect to the different angles of the walls through the advance and retreat of the elevation. The cross-dimensional horizontal table top, on the one hand, defines the division of the field. On the other hand, the semi-reflective texture of its metal material reaches a visual balance with the ratio of the vertical grating.

In the upper space, three blocks for usage are used by combining two volumes. In the public area, the volume departs from the glass panels of the outer walls. The line of sight is extended and a flow of light, shadow, and air with the materials is created. In the private area, by using mirrors, the volume pressure is reduced from the walls and reflect an extended space.

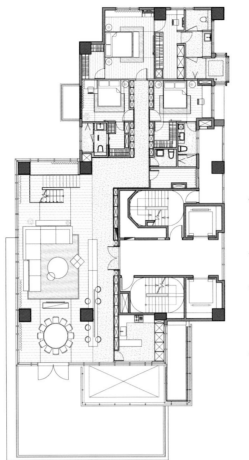

一层平面图 / First Floor Plan

二层平面图 / Second Floor Plan

艺术地生活，诗意地栖居
Live Artistic, Dwell Poetic

项目名称：威海吕宅
设计公司：飞形设计
设计师：耿治国

Project Name: Lv Mansion
Design Company: Office For Flying Architectur
Designer: Geng Zhiguo

凭栏眺望一片渤海水域，海平面上落日余晖，星芒四射，而万顷沧溟在此日昼被谪降时分，也缓缓从风动时的潋滟波光转换成一种曲线式涟漪。若从耸峙崖岸上的别墅露台近瞰，还能看见碎涛击荡礁石，情态万千。然而此滨海住宅原有建筑比较封闭，门窗狭小，且未考虑与周边环境融洽共存。因此，设计师耿治国为发挥基地坐拥的极致海景，以"打开盒子"的概念重塑三层住宅量体，不仅获得开放性极佳的空间视野，同时也顺应海岸线地形的起伏，让每个房间都能欣赏到极致的海景。当建筑外墙呈现在夕照下，感觉质朴且有一份柔和神仪。立于露台远望静水，心旷神怡。

本案业主是一位台湾儒商，对当代艺术，尤其是当代华人艺术有着极大的热情，他希望通过设计师巧妙的设计手腕，将本案打造成一个艺术收藏者之家，满足度假、会友、收藏艺术品的需求。因此，设计师把一楼辟为客厅、餐厅兼备会议功能的茶室，二层是主人的书房和卧室，客房则位于地下室，所有空间在具备了生活起居的必要功能的同时，也与艺术进行着对话。大面积的落地窗，使室内与户外环境互相渗透，让空间更加开阔，采光通风效果俱佳。

"无功能不设计，无艺术不生活"，空间设计的目的和基础是解决使用者的功能需求，而艺术则有利于提升生活品质。在室内开放的大空间的设计上，设计师以深色为主调，将"艺术"与"生活"相结合，既满足了生活起居，又使空间成为艺术展示的背景，实现艺术与空间及生活形态的对话。陈设艺术品是本案设计的一个重要的元素，在这里，它不再是让人仰望、顶礼膜拜的欣赏品，而是日常生活的一部分，唤起人们内心的艺术情怀.诸如遍布各功能空间的字画、抽象的画面、神奇的线条、或斑斓或黑白的色彩，在灯光的衬托下，美化了生活空间，增强了空间的美感和趣味性，带给人们无限的遐想。当然，这些字画（艺术品），也体现出业主高尚的生活情操以及对艺术品的热爱。

此外，光线也是本案设计的一个亮点。设计师把自然光线、生活起居的灯光、照射艺术品的灯光，通过调控，实现了其在空间中多层次的融合。

地下层平面图 / Basement Plan

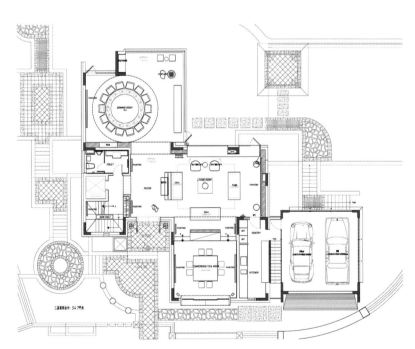

一层平面图 / First Floor Plan

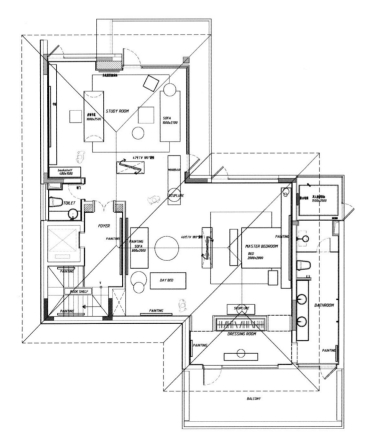

二层平面图 / Second Floor Plan

Across the railing against the Bohai sea, comes waters and waves. In a setting moisture flashed, even stars blur their glister, and boundless expanses of water are constantly changing their ripples. Bird views from the villa get a glimpse of waves which are hitting reef and immediately are broken into pieces. A project this space is with a relatively closed area, and narrow windows and doors, where to employ concept of opening door to reshape the 3-storey volume. The result is that the space is not opened to the exterior but well consistent with the geography, so each room can maximize its water view. During hours of sunset, the embracing walls becomes more pristine and much clearer. View on terrace can reach as far as where the water ends, breeze coming relaxing and rejuvenating.

The owner, a Confucian businessman from Taiwan, has a flaming enthusiasm for modern art pieces, particularly those by contemporary artists. He aims to build up a space for collection, family resorting and friends gathering. So the first floor is for the living room, the dining room, the tea room which can also serve as a meeting room, while the

second for the master study and the bedroom, and the basement for guest room. All sections initiate a dialogue with art pieces in meeting demands of life. Landing windows of large area maximize the communication external and internal, the space thus becoming broader and more convenient for daylighting and ventilating.

Functions cannot go without design, nor does life without art. Spatial design is aimed at and based on practical functions, and art is beneficial to promote life quality. As for the large area in the interior, dark hue plays a dominating role in a setting where to combine art and life. This not only meets living needs, and transfers the space into a backdrop for art display. So art, space and life can communicate with each other. Furnishing items, including accessories, as a prominent element, are not worshipped but find their way into daily life to evoke inner emotion of people. Take the paintings, works of Chinese calligraphy, lines and hues as examples. Paintings are in their abstract sense, lines are miracle, and hues are gorgeous or black-white. All against lighting, beautify the space in enhancing the spatial aesthetics and fun to generate imagination one after another. Of course, these art pieces express the personal love and individual life taste and sentiment.

Additionally, lighting plays another important role in the space, where daylight, lighting and lamplight to cast on art pieces make a multi combination with the aid of curtain or under control.

师法自然，木石有情
To Follow the Nature, Wood and Stone Become Affectionate

项目名称：台中湖心泊
设计公司：天坊室内计划有限公司
设计师：张清平
面积：180 m²

Project Name: Taichung Lake Center
Design Company: Tien Fun Design Planning Ltd.
Designer: Zhang Qingping
Area: 180 m²

在设计手法上，本案向树木学习谦逊，向石头学习坚毅，师法自然，延续以人文为本的理念，真诚创造家的永恒价值。行吟坐卧，都有天光相依，在慢节奏的生活里，听见欢快的心跳。

房子有价，氛围无价。空间装饰简洁素美，以一种极致精简的态度，将内涵深厚的人文观念体现于空间的每一个角落。在本案中，人文与绿意双轴并行，透过理性的机能规划，体现出乐山乐水的写意。阅尽繁华，回归家中，静坐于沙发，感受阳光倾洒进来的温暖、温馨情怀，油然而生。整体空间设计以黑、白、灰色调为主旋律，关系明朗，加上独特的造型与用材，共同塑造出一个张力强劲的生活空间。直线元素的运用，简洁纯粹，不张扬，不平淡，使人与空间有着一种较强的情感交流，给人们带来柔美、自由、随意的心理感受。遍布空间的石材，在设计师娴熟的设计技巧下，打散铺设，拼凑出独特、漂亮的图案。客厅电视墙上的干花装饰，质朴又时尚，尤为引人注目。

苏东坡有诗云："宁可食无肉，不可居无竹。无肉令人瘦，无竹令人俗。人瘦尚可肥，士俗不可医。"竹是高雅、纯洁、虚心、有节的精神文化象征，古今庭园几乎无园不竹，居而有竹，则幽篁拂窗，清气满院；竹影婆娑，姿态入画，碧叶经冬不凋，清秀而又潇洒。古往今来，"人生贵有胸中竹"已成了众多文人雅士的偏好，他们常借竹来表现自己清高脱俗的情趣，或作为自己品德的鉴戒。在卧室设计中，设计师巧用竹子墙纸修饰，不仅很好地提升了空间品质，还彰显出居室主人高雅的品位。本案专注家的温度，精心编制专属您的生活旋律。

平面图 / Site Plan

With the humility of trees and the perseverance of stones, the design of this project sticks to nature following to carry forward the humanity-oriented concept. The sincere guardian of home value creates a sustainable value for your home, where walking, sitting or lying can be accompanied with daylight in listening to your cheerful heart in a slow life.

House can be priced, but atmosphere not. The decoration for this space is simple and concise with an extreme attitude applied to material choosing. The humanistic connotation goes warm everywhere. Culture and green go parallel through rational planning function, reflecting a mood of liking hills and waters. At home, the comfortable sofa with warm sunshine coming through, an expectation of home and a desire for fine living are self-evident.

Black, white, and grey on the whole, the space is unified. Its clear relationship, unique shape and materials altogether bring forward a living space with a strong tension. Linear elements, simple and pure, low-key, not extravagant or dull, allow for a powerful emotional communication. That brings softness, freedom, rhythm and elasticity as well as speed, rigor and technology for people to feel. The stone around under the skilful hands put together a unique and beautiful design. Dried flowers on the TV wall of the living room are both plain and fashionable, so eye-catching.

As Su Dongpo, a great poet in North Song Dynasty, said that "dinner can go without meat, but dwelling cannot without bamboo. No meat diet would make people thin while no bamboo living tends to make people vulgar. Thin people can become fat, but people vulgar cannot be refined". As a symbol of nobility, purity and modesty, bamboo has since an internal part of garden and patio, which looks refreshing and beautiful throughout the year. Bamboo in heart has been sought by refined scholars. They often turn to bamboo to express their temperament and interest or taken bamboo as their example. And the bamboo-patterned wallpaper in the bedroom not only boasts the spatial quality, but sets off the personal taste. A project this space is that focused on home warmth, and is well-prepared and dedicated to your life melody.

隐在湖边，归在田园
Reclusive on Lakeshore, Returning to Countryside

项目名称：重庆中海黎香湖
设计师：琚宾
用材：桃花心木、米黄石、白玉石、钨钢、壁纸、铜
面积：650 m²

Project Name: CSCL (Chongqing) Lixiang Lake
Designer: Ju Bin
Materials: Mahogany, Marble, Tungsten Steel, Wallpaper, Copper
Area: 650 m²

桃花源是陶渊明笔下中国古代文人理想的隐居之所，世外桃源的"采菊东篱下，悠然见南山"的休闲生活，一直是回归内心东方美学的精神主导。

本案位于黎香湖这样一个现代桃花源。隐在湖边，归在田园。设计师用东方美学的六种心境设计出一个低调而温润的气质空间。

美心
设计师将自然的意境与当下的生活方式相结合，将文化元素融入到空间中。美学与自然和谐统一，形成静谧悠然的心境。

德心
"重回经典、回归传统"是中国美学所追求的方向。因此设计师把宋代文化意境作为本案的设计灵魂，着力营造一种"当代东方"的度假生活氛围。

创心
历史的情怀、时尚的气息使得这里散发出一种个性与共性、复杂与多元并存的"集古"气质。因此设计师将东方空间的气质美学与地域的矛盾性、文化的复杂性融合在项目设计中，并使之共融共生。

精心
精致的细节、雅致的氛围、目之所及或山色或湖景，让人沉浸在悠然的山居度假气氛之中。

艺心
本案以东方文化为出发点，运用装饰元素散发出美感，使空间呈现出一片蕙质兰心的东方风韵。时尚的软装搭配，融合传统纹样与精致的面料，形成现代触感。传统工艺与现代艺术相结合的饰品，透露出当代东方情怀。

匠心
精致的收口细节，木、石、金属的运用，精湛的制作工艺，东方设计独具匠心的细腻设计与西方工艺的巧夺天工被完美融合在本案之中，呈现出一个既是人文美学典范又具有国际化视野的空间。

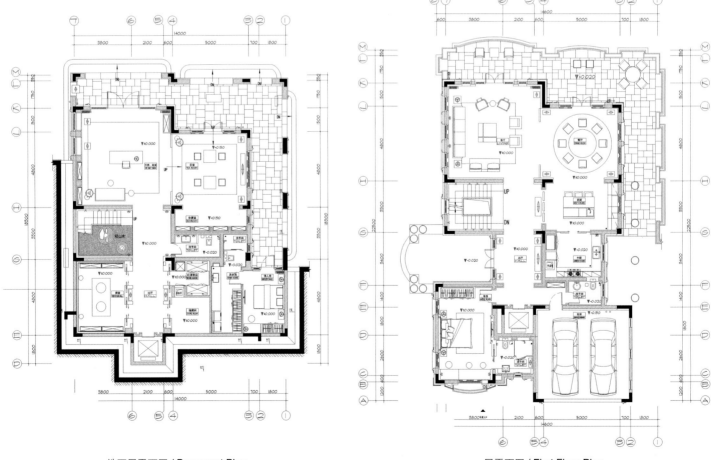

地下层平面图 / Basement Plan

一层平面图 / First Floor Plan

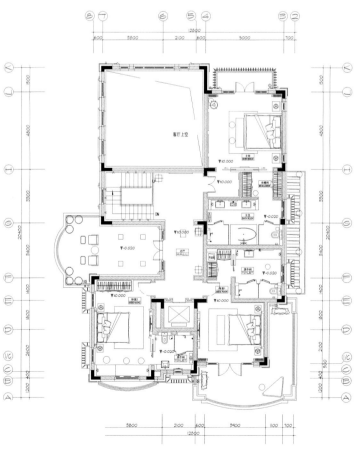

二层平面图 / Second Floor Plan

An idyllic land under description of Tao Qian (a famous Chinese poet in the 4th century) is envisaged to be an unsophisticated, pure and glorious fairyland. It has been regarded as an ideal getaway place for Chinese literati generation after generation. Despite the fast development of material world and the pursuit of luxury life instead of spiritual life in modern society, to find a way to a peaceful state of mind has been a main idea of oriental aesthetics.

A project this space is located in a modern setting of Peach Garden, where one can rest by the lakeshore and live an idyllic life. Six moods from oriental aesthetics are employed to express the mild and leisure spirits to the villa.

Beauty
Spiritual elements from culture and nature are withdrawn and fused into the modern lifestyle expecting to trigger the unison of aesthetics and nature and forge inner peace for people.

Excellence
Recurrence of classics and tradition is what Chinese aesthetic has been longing for. Therefore, the design was based on the spirit of culture in Song Dynasty. Under the concept of modern oriental vacation life, the design seeks common grounds for tradition and modern life.

Creation
The villa is a complex of commonality and individuality, multi-culture inherited and infiltrated history and the sense of fashion. Implicit oriental feelings are creatively infused into the design and united with the contradiction within regions and the complexity of the culture.

Elaboration
Exquisite in details and elegant in atmosphere, Lixiang Lake project provides people leisurely vacation with its beautiful landscape and fine design.

Art

Based on oriental culture, every accessory in the space expressed an elegant temperament on spiritual level and enhanced the modern feelings of space with fashionable decoration, traditional patterns and well-crafted materials.

Ingenuity

All exquisitely designed details, the use of wood, stone and metal in design, and the fine craftsmanship of the west have been unified within, unconsciously revealing a cultural space with oriental aesthetics and equal international view.

执笔缱绻书古意，
尽将流光付闲情

Writing Is Attached to a Description of Ancient Conception;
Time Can Be Spent at Leisure

项目名称：中星红庐65号别墅
设计公司：鼎族设计
设计师：吴军宏
摄影师：三像摄建筑摄影机构 张静
面积：900 m²

Project Name: No. 65 Red Villa
Design Company: Prosperous Clan Adorn Design
Designer: Wu Junhong
Photographer: Threeimages Zhang Jing
Area: 900 m²

本案定位为中式风格的度假别墅，宁静祥和的氛围中弥漫着浓郁的人文气息。满室的自然之意透露着中式的古朴典雅，予人无限惬意。推门而入，玄关背景墙后的公共区域以两列纵向排开的圆柱分隔，小客厅与餐厅则与半开放式的栅栏遥遥相对，三角梁样式的吊顶让人仿佛置身于老式住宅之中。统一的屋檐下呈现出来的大家庭氛围，使得屋主一家无论是喝茶还是用餐，都备感温馨。整个空间设计，古色古香。在格局和功能上，以现代生活为蓝本，突出度假和休闲的主题。玄关左侧的区域是客厅，往里深入，是摆放着文房四宝的书房。玄关的右侧，是连接地下室和二楼的旋转楼梯。考虑到老人的行动不便，设计师贴心地把老人套房安置在一楼。二楼是包括起居室、主卧室以及两间客房的私人空间。地下室则是丰富的公共活动区，有收藏室、茶室、影音厅以及SPA等休闲区域。

为了平衡木材带来的沉稳与庄重感，也为了注入几分现代气息，设计师采用了深色镜面来增加虚拟空间，制造出多层次的空间感。穿行室内，仿若行走在曲折、幽邃的小径，别有意境。首层走廊间，在大理石地面及深色镜面天花、墙面的默契配合下，整条走廊深远而悠长。地下茶室内，镜面天花则将展示柜内的陈列品无限延展，既扩大了室内空间，也丰富了空间语言。设计师在格局和功能上以现代生活为蓝本，最终打造出一个亦古亦今的休闲生活空间。

A project this space is positioned as Chinese-style villa for holiday, where the quiet and peaceful atmosphere is rich in culture. Natural, simple and elegant on the whole, the space allows for unlimited comfort. Behind the entrance wall two lines of column are lined vertically in the public area, which serve as partition. The small living room and the dining room face each other across a distance in a form of semi-open fence. The triangular-beamed ceiling makes people feel as if they were in an old house. The whole family under the same roof feel especially warm either drinking tea or having a meal.

Throughout the whole space is carried out the antique Chinese style while in terms of pattern and function, modern life functions as the blueprint by highlighting the holiday and leisure theme. To the left of entrance is the living room, and then the study where to contain the scholar's four jewels of brush pen, ink, paper and ink stone. To the right is the spiral staircase that extends from the basement to the second floor. On the second floor are family room, master bedroom and two guest rooms. The first floor is considerate and thoughtful to accommodate suite for the elder. Public activity naturally occurs in the basement, where there are collection room, tea room, audio-visual room, SPA and other recreational areas.

In order to balance the sense of the staid and the solemn, dark mirror is used to increase the virtual space and create a multi-level sense of space in injecting a bit modern sense. The meander throughout is like walking along twists and turns. With marble floor and dark mirror in the ceiling and walls, the hallway of the first floor looks far-reaching and long. In the tea room of the basement, the ceiling mirror infinitely extends the exhibition in the cabinet, expanding the interior space and enriching the space of language. With structure and function based on modern, a leisure living space comes ancient and modern.

以意带形,提纯人居美学

With Connotation to Promote Physical Shape, Dwelling Aesthetics Is Upgraded

项目名称:重庆黎香湖低碳社区 D 户型

用材:西班牙米黄、珊瑚红、帝皇金、爵士白、红色烤漆板、香槟色烤漆板、钛金条

面积:509 m²

Project Name: Lixiang Lake Show Flat D, Chongqing
Materials: Marble, Stove Varnished Board, Titanium Sheet
Area: 509 m²

本案主要针对高端人士,尤其是女性客户而设计,以"岛居纯独栋"、"完美星期天"为主亮点,打造出功能性强、风格时尚、柔美、温馨的完美居室,使之成为新贵阶层的第一居所。

设计以意带形,无论是空间造型还是家具设计均体现了西方建筑的架构体块感,于简单的形式中流露出淡淡的写意情境。面积大、比例精巧、不同立面的材质处理,自然和谐,不着痕迹地重新定义出空间的布局和结构。

装饰元素体现出装饰艺术的细节。设计师从西方生活中提取精致的细节,以上乘的品质、时尚的气息、优雅的风度来诠释空间。家具和地毯的纹理细腻,丰富了空间质感。颜色搭配以红、白、黑三种经典色彩为主,红得热情妖娆,黑得优雅动人,白得纯净无瑕,形塑出大气、时尚的空间气度。而或风雅或时尚的饰品恰当地散落在居室的不同角落里,奏响生活的主旋律。

对于平面的处理,设计优化了平面户型并梳理了功能流线。整个空间将钛金属的装饰线条运用于不同材料的表面。在天花的细节造型上,又结合了灯带和射灯的基础照明,让灯具与天花造型达到统一和谐的效果。立面的金属条与烤漆板利用激光切割等新技术,保证了细节效果。此外,整个空间的细节收口采用了隐性的收口方式。

地下层平面图 / Basement Plan

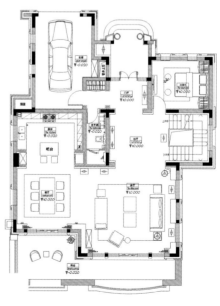

一层平面图 / First Floor Plan

This project targets high-end people as its consumers, particular woman group. The appealing point of "island living, single building "perfect sunday" together with strong functions, fashion and perfect interior makes it the best choice for upstarts.

Design stresses that spatial connotation can shape the physical appearance. All of spatial modeling and furnishings embody structural block of the western building, revealing an enjoyable state in a simple form. Of large size, ratio accuracy, and different material, facade redefines the spatial layout and structure but leaving no trace in a natural and harmonious approach.

Decoration indicates details of art deco. Essence of western life with its upper quality, fashion and grace interprets the space. Furniture and carpet involves more subtle texture of material to enrich the space. Red enthusiastic and enchanting, black graceful and moving, and white pure and flawless, all makes a sense fashionable but still of grandeur and generosity. Accessories elegant and up-to-date everywhere serve as the theme of the space.

As for the plan disposal, the original plane has been optimized and functions have been put into order. Decorative sheets of titanium are employed onto surfaces. The use of lighting belt and spot lights for the ceiling unites both in terms of shape to achieve a harmonious effect. New techniques like metal sheets on the façade and laser-cut stone varnished boards guarantee a detailed effect. Additionally, a recessive close-up approach is employed for the spatial detail.

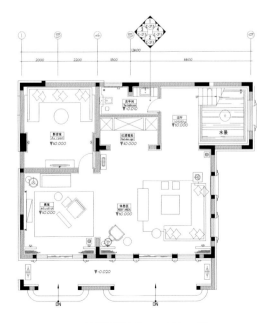

二层平面图 / Second Floor Plan

浸染诗意的家园
Dyed in a Poetic Home

项目名称：绿地·海域观园样板间
设计师：胡斐
摄影师：三像摄建筑摄影机构 张静

Project Name: Greenland Sea-viewing Show Flat
Designer: Hu Fei
Photographer: Threeimages Zhang Jing

如同在中国的水墨画中浸染过一般，黑白灰的底调描绘出中式生活的意境。推开建筑的双入户原木大门，即可看到含蓄经典的现代亚洲风格内装，将"出则含蓄典雅，入则舒适阔绰"的中式住宅精髓发挥得淋漓尽致。户内规划复合五重院落空间，引自然入室。其中，独有的中庭花园，用自然景致装点客厅与餐厅。二层、三层卧室套房不仅附带观景露台，更有超大屋顶花园。附赠的超大地下室，享尽视听之娱，直通地下独立车库，实现人车分流。

客厅宽大、明亮、多面采光，中西合璧，既有传统的中式家具，又有宽大舒适的西式沙发；既有中国结等装饰品，又有颇具结构感的欧式壁炉。空间整体色调古朴，就像一幅笔墨浓重的中国画，充分展现出中国特有的文化底蕴以及居者不凡的审美眼光。吊顶处又适当地带上一笔辉煌的金色，锦上添花，提升了空间的品质，展现出空间的华贵与夺目。"雪映红梅梢头笑，落枝近地红雪飘"，倚靠主墙的那一抹红与白的交织，最为引人注目，画面浸染白雪皑皑，红梅花开，仿佛将人带入了"踏雪赏梅"的浪漫意境，让人不由心生美感，爱到极致。

餐厅设计用色高贵，彰显出主人的高贵品位。金属质感的餐桌椅，完美的线条、铮亮的颜色与中式古朴的木质吊顶形成鲜明的对比，无论从颜色还是从样式上，都将现代与古朴相结合，时代感十足。

不同的卧室定义不同的主题，或浓烈或隽永，都让人眼前一亮。不论是一只陶罐配一束鲜花，还是舒适的软床、布艺、精美的窗花雕饰都让人沉醉，突出一种安详、宁静的氛围。墙壁上一幕幕花卉图案，从单一的花朵到盛开的繁花，自然气息扑面而来，使居室呈现出不同的风情。

样板间还设计有品茶室、书法室与收藏室，利用茶具、折扇、瓷器、窗格、中式家具等具有中式古典气质的物品渲染出浓郁、古朴、具有中式味道的意境，如水墨画般，有虚有实，有明有暗，儒雅沉稳。每件家具、饰品的搭配与摆放都仿如有一个故事，透着一种淡淡的古典怀旧情怀。

This is a project that is like being dyed in a Chinese ink painting, whose three primary colors make a vivid account of a Chinese style life. Across the courtyard, a double-door entrance leads into the interior of Asian style, reserved and classical, where dwelling essence of a Chinese house is exerted to the utmost, that the external is reserved and refined once out of home, while the interior must be comfort and broad. Five sections are planned in the internal space, where the nature is taken into. The atrium garden makes a natural patio, natural elements from which find their way into the living room and the dining room. Bedroom suites on the 2nd and 3rd floors are fixed with view balcony and super large roof garden. The basement, besides home theater, accommodates a double independent garage.

The living room takes in daylight from multi sides, making an east-west space in a large and bright setting. Chinese-styled furniture is accompanied with large and comfortable western sofa. Accessories like Chinese knot contrast with the European structural fireplace. Primitive on the whole, the space looks like a painting of Chinese ink. Its elegance and harmony fully reveals the culture accumulated in the history of China and the special aesthetics of the occupants. The gold patch in the suspended ceiling makes the cherry on the cake, enhancing the spatial quality and indicating the luxury and glister. Plum becomes much redder against the white wall; branches are falling almost to touch the ground with snow floating. The red and the white on the main wall are so eye-catching, an image that has already led into a romantic artistic concept of appreciating plum in snow. Aesthetics are catalyzed within to make a paramount love.

Hue nobility in the dining room sets off the personal refined taste. With shiny hues and line perfectness, Metallic dining table and chairs make a sharp but intended contrast with Chinese primitive suspended wood ceiling.

Different bedrooms are designed with changing themes. Some are strong while some are meaningful. All, however, are enlightening. Making people lose, a pottery, a bunch of lowers, a soft bed, fabric and window decoration are intended for a peaceful and quiet atmosphere. Flowers single or clustered and plants on the wall, are blowing a natural air to present different style in bedroom.

Additionally the show flat houses a tea room, and spaces for making Chinese calligraphy and collection. Tea ware, folding fan, porcelain, and window of lattice and furniture of Chinese style render an artistic concept rich, and simple and unsophisticated, like that of Chinese ink painting. The empty goes with the solid, the bright with the dark and the learned and refined with the staid. All items of furnishings and accessories seem to be telling of a style, confiding a touch of warmth classical and nostalgic.

新传统文化视野下的精致文化生活韵味

Fine Culture and Charming Life by New Traditional Perspective

项目名称：中海文华熙岸
设计师：黎广浓、唐列平

Project Name: CSCL Culture and Prosperity
Designer: KenLai, LPTong

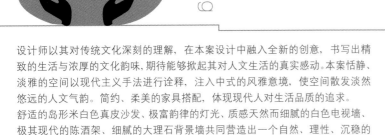

设计师以其对传统文化深刻的理解，在本案设计中融入全新的创意，书写出精致的生活与浓厚的文化韵味，期待能够掀起其对人文生活的真实感动。本案恬静、淡雅的空间以现代主义手法进行诠释，注入中式的风雅意境，使空间散发淡然悠远的人文气韵。简约、柔美的家具搭配，体现现代人对生活品质的追求。
舒适的岛形米白色真皮沙发、极富韵律的灯光、质感天然而细腻的白色电视墙、极其现代的陈酒架、细腻的大理石背景墙共同营造出一个自然、理性、沉稳的空间。

为塑造一个张力感极强的生活空间，客厅整体色彩浓淡相宜，电视背景墙以马赛克拼贴装饰，搭配白色长条的大理石电视台面，极具现代感和人文气息。中式实木窗棂的运用更显端庄稳健，汉白玉摆饰的运用，无不彰显出现代的东方文化，使整体空间大而不空、厚而不重，有格调又不显压抑。茶厅墙面采用裸露的灰砖堆砌，中式的木窗格门扇与案几，天然树根雕凿的茶台，像木雕工艺品一般，既具有实用价值，又有鉴赏价值，古意甚浓。
主人房漆白的木织柜门，优雅大方。实木格栅装饰屏风，古朴的窗棂格纹，兼

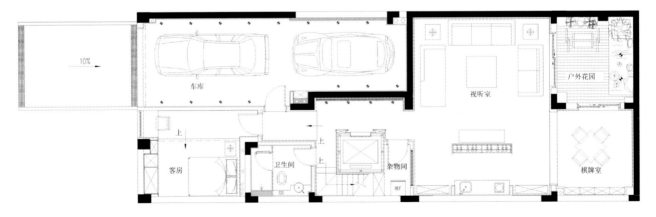

地下层平面图 / Basement Plan

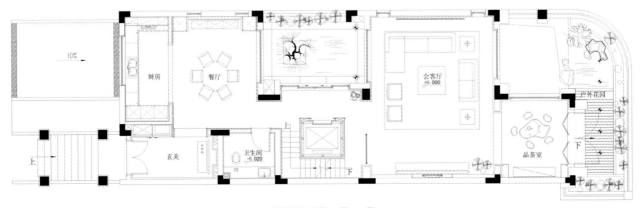

一层平面图 / First Floor Plan

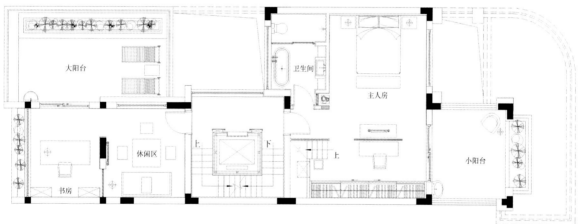

二层平面图 / Second Floor Plan

三层平面图 / Third Floor Plan

具古今；轻盈曼妙、亦虚亦实，灯光柔和宜人；床头亦做成白色实木格栅背景墙面，与屏风互相呼应；地面柚木地板，古典高雅；墙面大片的白色凹凸石面，简洁纯朴，旨在营造舒适的情趣生活。空间动静相宜，轻掀窗帘，美景尽收眼底。客房简朴优美、内敛淡雅，追求一种宁静休闲的生活境界。书房极简中式的家具，稳重大方，灯光强弱相宜，给主人以宁静致远的心境。卫生间功能实用美观，演绎不一样的品质生活。

From the moment, our understanding and perception would be used to start an all-new creation and share to make an account of fine life and exquisite culture. This is aimed to stimulate a true desire for cultural life. This project makes a calm and refined space with modern approaches to implant a Chinese-styled mood, so the space can have a cultural temperament that can last indiffenet bur very long. Furnishings and accessories aesthetic, simple and soft fit in well with modern pursuit of life quality.

The ice-cream leather sofa is comfort and shaped into form of island. The lighting is very dynamic. The white TV backdrop is of natural and delicate texture, which becomes more natural, reasonable and staid when combined with modern wine rack.

In order to make a living space with a strong sense of tension, the living room is wrapped in hues that's in a proper ratio. The TV backdrop is embellished with mosaic, whose white long marble surface is quite modern and cultural. The Chinese window lattice compliments the elegant and staid of such a section. All with white marble sketch out a modern eastern culture, the space thus becoming solid even when large, not heavy even when thick, and not oppressive even when graceful. Walls of the tea room is endowed with feelings done by accumulating grey bricks. Door and table are in form of window lattice. Tea table is carved out of a whole tree root, like art pieces of wood carving. That's both practical and of great value in making a very strong sense of ancient conception.

In the master bedroom, the cabinet door is of wood and painted white. It's elegant and magnificent, skillful and tolerate. The grating screen is of solid wood and the window lattice is of primitive simplicity. The lighting is soft and pleasant. The white bedhead is of solid wood grilling, echoing with the screen. The flooring of solid teak is classical and refined. The wall coated in large concave-convex stone, is white, simple and primitive to make a comfort life. Space staid and dynamic, the window takes in a panoramic view. Simple and primitive, guest rooms are conservative but graceful in search for a life setting of peace. The study is elegant and refined, where Chinese-styled furnishings are minimal, steady and magnificent, lighting whether strong or weak is proper and appropriate, and you feel nothing but that you can accomplish something by leading quite a simple life. The bathroom is practical but very beautiful, presenting a qualified life of another kind.

京华烟云
Moment in Peking

项目名称：沿海地产赛洛城样板间
设计公司：重庆品辰设计
设计师：庞一飞、张雁
用材：藏青色木作、白色软包、木雕花
面积：125 m²

Project Name: Show Flat of Silo City
Design Company: Pinchen Design
Designer: Pang Yifei, Zhang Yan
Materials: Carpentry, Upholstering, Wood Carving
Area: 125 m²

京华烟云般的民国情致，古典的中式元素配以现代的时尚功能，构成了本案的主题风格。在这里，岁月静好，时光温柔，听一曲古乐，品一杯香茗，轻握杯盏，闭上眼睛轻轻地印下蔻丹唇印，寄托心头的一缕缱绻情思。阳光迎窗，空间述说着缤纷美丽。万籁俱寂、清风徐来，在花开的时间里相聚相守。空间如文、极具灵性，给予人们以诗的意境及画的意象。

穿越夜的弥漫，依稀浮现，如梦似幻。窗外又是这个季节，春日添细雨。记忆镌刻在时光岁月的洪流中，清浅流连，彼此相依，雾积云散，行到水穷处，坐看云起时。

袅袅娜娜的佳人轻步走到梳妆台前，台上圆润而富有光泽的珍珠，璀璨夺目的耳饰，衬托出女子的高贵典雅与如兰气质。望着镜中美丽的容颜，女子手执木梳，临窗晨妆。温润的木齿在油光水滑的丝缕中穿行，自上而下滑溜而落。扬手间，宽大的衣袖轻轻滑下，露出半截洁白如玉的手臂。"琉璃梳子抚青丝、画心牵肠痴不痴"，思念如蜜。这是关于她，一个幸福女子的故事。

如果思念是一种等待和期盼，你又会选择在怎样的场景中等待、静静地守候心中的梦，本案这样带着民国情愫的空间会是一个不错的选择。中式的花鸟漆画、柜子上如同金箔的亭台楼阁造型、酷似汉朝说唱俑的配件……徜徉其间，犹如翻开一部厚重的线装书，处处散发出氤氲的人文气息。复古中透着现代风格的餐厅，将古典的悠然与现代的线条巧妙融合，既现代又不失古韵。休闲区的书法，更增古雅韵味。卧室清雅、干净，没有多余的装饰，却留下一墙似锦繁花。卫生间石膏线做出的中式回字格装饰，更好地体现了中式风格的韵味。

一切都是最好的安排，似在诉说一个古雅、美丽的传奇，让人久久地沉浸其中、难以抽离。

The temperament and interest descripted in Moments of Peking, and classical Chinese elements with modern fashion make the feature of this project. Here presents time of peace and quiet, where you mind would wonder off as you want with tea in hand and music in ear. Sunlight through the window is telling of the spatial beauty. Breeze comes gentle in an absolute tranquility. The space is poetic and spiritual, giving people a poetic mood and painting imagery.

Nocturnal softness rises vaguely, like a fantastic dream. Another spring it makes now with raining flowing away but memory into. Streams are going linked and woven with fog cluttered but cloud disappearing. It's here that you can get into a mood with which you like to explore where the water comes to an end and you tend to idle your time away with cloud flying.

On the dresser there are glistering and solid pearls and ear wear to set off the nobility of the female. Looking into the mirror, a beauty, with a wooden comb up and down her hair, makes a scene of peace and quiet. With holding-comb hand moving, the marble arm in the large gusset completes a happy story of the hostess.

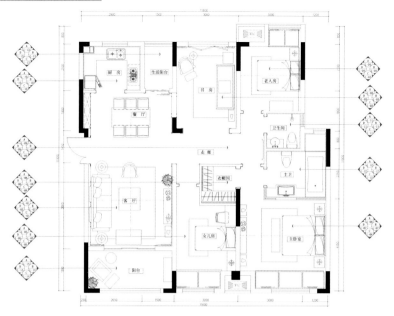

平面图 / Site Plan

If missing means waiting and expecting, then what scene can be complimentary that you would prefer for your always dream? That it is. Here it is. The project of this space with a good description of the Republic of China completes a good choice. Lacquer painting of flower and bird, cabinets with gold-foil-like pavilions, and singing tomb figure which seems to have gone though from Han Dynasty, make you feel as if you were opening a thread-stitched book. The dining room retro and modern makes a good combination of the classical leisure and the contemporary lines. Works of Chinese calligraphy add more ancient flavor into the leisure area. Bedroom is elegant and simple without surplus decoration but with a whole wall embellished with a mass of flowers. The Chinese-style fret with gypsum better embodies the taste of Chinese style.

Everything is best arranged. It seems to tell a quaint, beautiful legend, where people have long been immersed, unwilling to be kept away.

心闲不染尘，家藏笔墨香
A Heart at Leisure Is Free of Dust; Brush Pen and Ink at Home Are Fragrant

项目名称：中洲中央公园二期 11-B01 样板房

设计公司：KSL 设计事务所

设计师：林冠成

用材：白砂石、橡木、黑檀木纹、玉石、木地板、烤漆板、墙纸、

地点：广东深圳

面积：148 m²

Project Name: Zhongzhou Center Park Show Flat 11-B01

Design Company: KSL Design Company

Designer: Lin Guancheng

Materials: Gravel, Oak, Veneer, Stone, Wood Flooring, Stone Varnished Board, Wallpaper, Leather Sheet

Location: Shenzhen, Guangdong

Area: 148 m²

本案采用的新中式设计风格，不仅是对传统的继承，更是将传统精神理念融合于现代都市生活，通过提取传统中式设计的精华元素和生活符号，以合理的搭配、布局、设计，将传统中式的美学韵味悄然融入现代人的居住体验之中。严谨有度的设计语言完美诠释着品质高洁的精神气度，无论色调把握还是氛围营造都强调古典与现代的结合，传统与时尚的碰撞，整个设计典雅而温润，稳重而安详。

客厅在色彩搭配上下足了功夫，将黑色的稳重、木色的纯净、藏蓝的冷静与银色的高贵融于整个空间，直线与分割的运用让空间充满传统理趣，菱格屏风用于电视背景和空间分割充满创意，墙壁以如椽大笔装饰，配以丝绸抱枕和紫砂茶具平添了多重文化趣味，整个空间规整大气，品位高远。

原木色餐桌保留了明清家具的清雅气度，藏蓝与白瓷餐具营造宁静的就餐环境，一束明黄的迎春花开出满室的烂漫蓬勃。餐厅与厨房连通，巧妙地用黑色屏风将小吧台分割开来，趣味横生。

书房以黑茶色为主色调，营造出古朴安静的书香氛围，简洁造型的案桌和博古架弥漫着浓郁的历史气息，简洁而有力度的设计提升了空间的格调。

卧室色调雅致，空间极为舒展，通体木纹饰面的墙壁搭配黑色菱格让人心神宁静。飘窗的设计极为精致，两盏古典造型的灯饰仿佛穿越了时空，倚窗而坐、品一杯香茗，沉静而优雅。而另一间紧贴整个墙面的收纳柜兼具了收纳与展示的功能，实用性与装饰性完美结合，造型规整，设计精巧。

The neo-classical style for this project is not to carry forward the tradition but also blend tradition into modern city life. By extracting essence and life symbols of Chinese style which then undertake match, layout and design, Chinese aesthetics has been fused into modern dwelling place. Rigid and perfect design interprets a qualified spirit. Whether hue or air stresses the combination of the classical and the modern. With the collision of the tradition and the fashion, the whole space is warm, staid and peaceful.

The living room pays attention to nothing but color match by fusing into the entire space black stable, white pure, dark blue calm and silver noble. The use of lines and partition makes the space filled with tradition and interest. The screen for the TV background and space division allows for spatial creativity. Large pens decorate walls like rafters. Silk pillows and purple sand tea sets bring forward multi-cultural tastes. The entire space is really neat, crisp and lofty.

In the dining room, the wood color table retains the elegant furniture of Ming and Qing Dynasty bears. Porcelain tableware of dark blue and white creates a peaceful dining environment. A bunch of bright yellow flowers bloom flourish out. Linked together, the dining room and the kitchen are separated with black screen and small bar.

Dark brown is primarily used in the study to create a quaint, quiet scholarly atmosphere. Simple long table and antique-and-curio shelf are rich in historical flavor, plain but powerful to enhance the intensity of the style space.

As for the bedroom. Bedroom 1 is of elegant tone and extremely comfort. Walls are coated in wood finishes and black diamond of lattice. Bay window is pretty delicate. Two classical lamps feel as if through time and space. A tea by the window is so quiet and elegant. In bedroom 2, the storage cabinet close to the entire wall can also be used for display. Practical and decorative, it is sophisticated and styling.

阅尽千帆，情系东方
View of Sails, Passion for East

项目名称：万科松湖中心别墅
设计公司：深圳市派尚环境艺术设计有限公司
用材：天然大理石、天然木饰面、绣花皮革、艺术玻璃、拉丝铜
面积：500 m²

Project Name: Vanke Songhu Center Villa
Design Company: Shenzhen Panshine Interior Design Co., Ltd.
Materials: Marble, Veneer, Leather Embroidery, Art Glass, Brushed Copper
Area: 500 m²

项目位于万科松湖中心——松山湖沟谷公园内，兼具会馆功能的500平方米商务别墅，特为阅尽千帆的成功企业家打造。

这一类人群兼具国际化的视野和民族文化自豪感。对西方化的生活持开放和包容，甚至依恋的态度，但同时也对东方美学所独有的含蓄和隽永饱含着深情。中西合璧的设计应该是适用于此类空间的方案，但是在设计中，设计师并没有不假思索地沿用一些中西方文化中的符号性元素。对于传统的元素，设计师经过了感性的提取，理性的加工与运用，保持了空间的时代性和独特性。例如，传统中式纹样经过再设计，被运用到了皮革、艺术玻璃、镶板等空间细部中，另一方面，通过现代的表现形式营造出符合现代人生活的空间环境，让人们在内心得到熏陶后再从容感悟传统的魅力，境由心生，在不经意中感受中西文化

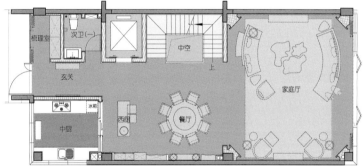

一层平面图 / First Floor Plan

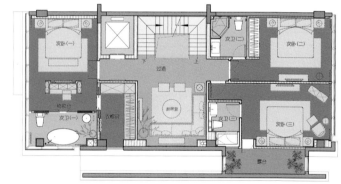

二层平面图 / Second Floor Plan

各自独特的魅力。

空间运用了三种深浅不同的石材，以自然界各种纯天然的纹理及材质，在视觉、触觉上丰富人们对空间的体验与感受。木与石的纹理之美，以及玻璃的清透质感，在光影下呈现出繁复不规则的变化，透出典雅的氛围。

客厅的空间处理中，既不失仪式感，又具备很强的自由度。在空间细部的处理上，设计师着意于素材的质地与对比，香槟金色的镜钢点亮了空间，展现出空间的层次感。

空间以中性的米色调和灰调为主色调，具有原木色泽和纹理的木饰面，自然地传达出温暖舒适的感觉，很好的烘托了沉静安宁的空间氛围，与窗外的自然景观和室内木雕装置仿若浑然天成。

地下层水景别出心裁的凹凸造型，使得水流在落下的过程中，编织出织锦般的美妙纹理，在庭院休息平台上观景的人，不仅可以饱览周边自然的美景，也可以悠闲欣赏自家庭院中的妙趣横生的园林小品。

Located in the center of Songshan Lake Park, the space is a commercial villa with an area of 500 square meters, only for entrepreneurs who have survived fierce competition.

A group the targeted consumers is who has an international perspective and a national cultural, proud, inclusive or rather attached to westernized life on an open, but full of affection for unique subtle aesthetics and

connotation of the eastern aesthetics.

Chinese and Western design should be suitable for such a space, but the practice for this space has not unthinkingly followed symbols of Western culture elements. Traditional elements have gone through emotional extraction, and under reasonable processing to keep epochal character and peculiarity. For example, traditional Chinese patterns undertake re-design and are applied to the leather, art glass and paneling. On the other hand, by creating a modern manifestation of the space environment that is in line with the modern life, people consequently get nurtured in his heart and then perceive tradition at leisure, in which course the charm of Chinese and Western culture can be experienced inadvertently.

Three stones in different shades are mainly used as well as various natural textures and materials to enrich people's experience visually and tactilely. The beauty of wood and stone, and the transparency of glass show complex irregular variations with light cast on, revealing an elegant atmosphere.

The spatial processing in the living room has both a sense of ritual and a strong degree of freedom. Inked texture and contrast material are deliberately made. The use of champagne gold mirror lights up the whole space, whereby completing a sense of space level.

In a neutral setting of grey and beige, veneers of wood texture and finishes, naturally convey warm and comfortable feeling. This well heightens the quiet, peaceful space atmosphere. With the natural landscape and the interior wood carving an image that everything seems to be born in the nature.

Waterscape in the basement is of ingenious convex-concave shape, water falling and making a tapestry of wonderful textures. The rest on the patio platform not only can enjoy a surrounding natural beauty, but also a leisurely garden with small ornaments of their own.

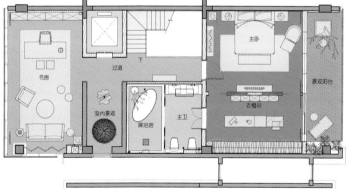

三层平面图 / Third Floor Plan

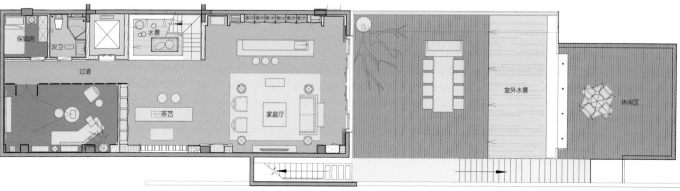

地下层平面图 / Basement Plan

水墨淡雅，回味悠长
Chinese Ink Painting Is Quite Refined and Worth Aftertastes

项目名称：泉州鼎盛大观样板房 A1 户型
设计公司：福建国广一叶建筑装饰设计工程有限公司
设计师：陈世忠、邱屹梅
设计审定：叶斌
摄影师：施凯
用材：大理石、黑钛、墙纸、软包
面积：98 m²

Project Name: A1 Show Flat of Grand View in Quanzhou
Design Company: Fujian Guo Guang Yi Ye Decoration Design
Designer: Chen Shizhong, Qiu Yimei
Design Approval: Ye Bin
Photographer: Shi Kai
Materials: Marble, Titanium, Wallpaper, Upholstering
Area: 98 m²

本项目为样板房，空间被规划成三房二厅的实用格局。设计师为了让空间表现出年轻业主的个性特质，大量运用刚劲的线条及经典的灰白黑色调，打造出充满都会时尚气息的住宅样板间。为使空间洋溢年轻、大方的气氛，特地设置开放式厨房。灰色云石、黑钛与木纹的搭配，不仅提升了空间的层次感，也丰富视觉画面。

A Show flat this space is that is divided into a layout of three bedrooms and two halls. In order to express personality of the young owner, bold and vigorous lines and three primary colors are involved to create a fashion sense in a metropolis setting. To overspread an atmosphere young and grand, an open kitchen is fixed. Meanwhile, grey marble is used in large amounts with black titanium and wood grain, which enhances the spatial layer in enriching visual effect.

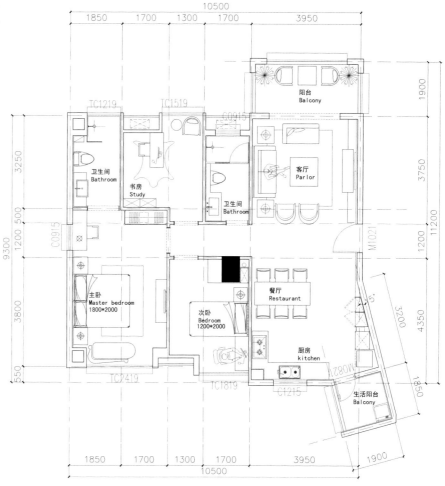

平面图 / Site Plan

一曲古韵，千年风雅
Ancient Rhythm, Everlasting Grace

项目名称：福清别墅
设计公司：福州恒观顾问设计有限公司
设计师：陈标

Design Company: Fu Qing Villa
Design Company: Fuzhou Heng Guan Consultant Co., Ltd.
Designer: Chen Biao

对于本案设计师来说，装饰空间的展示必须植根于中国传统文化的精粹，并吸纳当代的时尚元素，从而使整体空间弥漫着现代中国风的美妙、爽朗、隽秀与绚烂的气息。

本案建筑的外观与边缘在清朗月光的映射下愈发透亮，其线条的平衡与垂直、物面的凸起与凹入，都沐浴在柔和、静谧的光线之中，让人心旷神怡。

在室内设计中，设计师更是匠心独运，以鲜明的"中国红"为主要色调，间或辅以其他衬托性的颜色，架构出斑斓绚丽的空间画面。旋转式楼梯采用浅金色木板进行铺贴，木质扶手边角圆润，触感舒适、圆滑，给人非常安全、贴心的感觉。伞状圆形吊灯让光线从上往下弥散开来，给予这个空间温馨、柔和的光亮。走廊采用五块长方形大理石铺贴，间隔之中与两侧空隙均以大量鹅卵石进行填补，颇有于溪水中放置的踏脚石感觉，让简单的走廊也大放光彩，有了灵动、隽秀的质感。走廊一侧墙面大量铺贴浅黄色大理石，另一面与天花板均采用木质柱状装饰，将酒窖与走廊间隔开来。客厅窗棂造型的背景墙与柱子间的空隙也为该空间增添了通透性和光亮感。

茶室以淡茶色为背景色，选用暗茶色墙纸铺贴，突出墙面的层次感。屏风上洋洋洒洒一篇李白的《将进酒》，把茶室的古色古香与诗情画意表现得淋漓尽致。墙上一个"静"字，既是对人生的感悟，又彰显了中式的静思禅意。

最后，设计师运用"祥云"装饰吊顶，在木壁格子置放代表"丝绸之路"特有元素的瓷器，在廊道中安排木雕等工艺品，在玄关铺贴中国水墨荷花图案等，各种元素混搭配置，或清爽沉静，或浓墨重彩，展现了装饰手法上的丰富性与多样性，强调了设计是对于装饰空间的展示，是根植于中国传统文化精粹的观点，也使本案在装饰性与实用性的巧妙结合中得到艺术审美的愉悦。

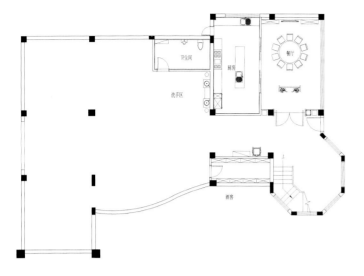

一层平面图 / First Floor Plan

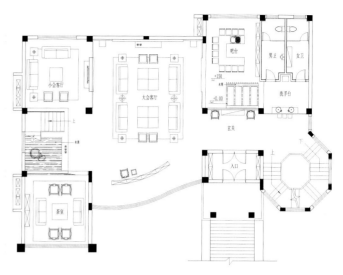

二层平面图 / Second Floor Plan

A decorative space of high quality must be rooted in essence of traditional Chinese culture for designer Mr. Chen Biao by withdrawing modern elements to overspread a Chinese style throughout that is marvelous, delicate and refined, bright and clear, and gorgeous.

Design for this space starts from the building's appearance, making a facade with its whole body reflected transparent. Lines are consequently balanced and vertical, surfaces concave-convex. All are bathed in light softness, peace and quiet to allow for inner pleasure and relaxation.

The interior design is exerted with ingenuity, where Chinese red is persistent with hues of other kinds here and there to offer a gorgeous scene. The spiral stairway is wrapped in light-gold veneer. Its handrail and corner are smooth, comfortable, safe and considerate. The umbrella-like droplight casts warmth and softness. Corridor is paved with five rectangle marble pieces, with amounts of pebbles filled in the gap, like stepping stone in the brook. A very simple design for the corridor has now become lither and beautiful. One wall in the corridor is coated in light-yellow marble, while the other embellished with

wood column decoration. The backdrop in the living room is shaped into window of lattice, enhancing transparency and lucency of the space. Walls in the tea room mainly involve color of green tea, with black-tea colored wallpaper to set off the layers of the space. The poem of *Invitation into Wine* by Li Bai, a great poet in Tang Dynasty brings out incisively and vividly the antique flavor and poetic illusion. The Chinese character of Jing, meaning peace and quiet, is the inspiration got as time goes by and sets off the Chinese Zen.

Finally, the suspended ceiling is patterned with auspicious clouds. Onto the grid of wood screen are porcelain pieces typical representative of Silk Road, in the foyer lotus images of Chinese ink painting. Elements of kinds are merged into, some refreshing, some peaceful, some plain and some colored to show a rich and variable decoration and stress the root of Chinese culture. The space is thus made unique, funny and interesting, providing artistic aesthetics in combining decoration and practicability.

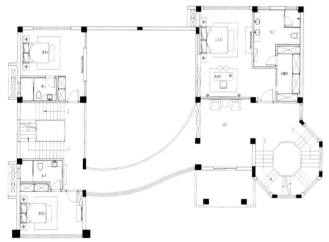

三层平面图 / Third Floor Plan

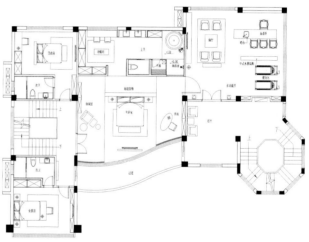

四层平面图 / Fourth Floor Plan

五层平面图 / Fifth Floor Plan

Feel the Eastern Zen Style — Living Space X

Feel the Eastern Zen Style — Living Space X

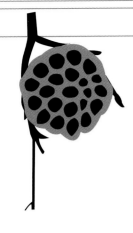

花将色不染，
身与心俱闲

Inner Peace and Quiet

项目名称：西安新兴·新庆坊
设计公司：5+2 设计（柏舍励创专属机构）
用材：爵士白大理石、意大利木纹大理石、白橡木饰面板、黑钢镜
面积：110 m²

Project Name: New Prosperity in Xi'an
Design Company: 5+2 Design (Perceptron Subsidiary)
Materials: Marble, White Oak Veneer, Black Stainless Steel
Area: 110 m²

本案为两厅三房两卫户型，风格定位为浅色调的现代中式。整个空间毫无吊顶装饰的白色顶面，剔除了多余的元素、色彩、形状和纹理。通过对空间的重新划分，使自然光线可以进入到客厅、餐厅、厨房，从而营造出素雅宁静、明亮舒适的居室氛围。客厅干净清爽的米色系纹理墙纸、典雅精致的木花格玻璃屏风及朴素柔和的卤素射灯，形成了现代中式的家居风格，给人以大气、素净、浑然一体的感觉。设计师通过对材料的极致运用，将餐厅的空间布局如山水画一样呈现出来，提升了整个空间的质感和画面感，营造素净、简约的空间氛围。主人房采用玻璃隔墙区分主卫与休息区，延伸空间视觉，增强空间的通透感；弧形的天花板、白色的柔毛地毯、浅黄色的台灯柔化了整个空间氛围，营造出温馨浪漫的气氛。孩子房的空间布置由床榻及书桌组合，暗藏灯带和木质家具的结合，既满足了现代生活的功能需要，又可以在感受自然简约的同时安享宁静温暖的家居氛围。客房通过利用四边方正的书桌、简练硬朗的线条、拉伸了整个空间，中式的木花格玻璃屏风与落地窗相呼应，让空间变得更平和、宁静。

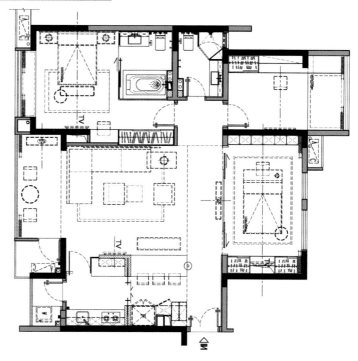

平面图 / Site Plan

A project of three bedrooms and two living rooms this space is that is designed with light color and Chinese style. The white ceiling not suspended keeps away all surplus elements, tones, shapes and textures. By redefining the space, daylight finds its easy access into the living room, the dining room and the kitchen, creating feelings bright, and comfort. In the living room, beige grain wallpaper looks neat and crisp, glass screen of wood grid is elegant and refined, and halogen spot light is simple and soft. All confide in a modern Chinese dwelling style, allowing for unified feelings of grandeur and generosity. Materials are used to the utmost in the dining room, whose layout looks like a painting of landscape, enhancing the texture and imagery of the whole space. And the air is thus simple, plain and neat. The child's room consists of bed and desk, with lighting belt and wood furniture combined. This not only meets needs of modern life, but also exposes people to home warmth and peace. The guest room is extended with a square desk of straight lines, where glass screen of wood grind echoes with the landing window. The space consequently becomes peaceful and quiet.

当时明月在，曾照彩云归

The Ever Bright Moon Escorts the Cloud to Return

项目名称：广州星河丹堤别墅
设计公司：矩阵纵横
用材：爵士白大理石、孔雀金大理石、新月亮古大理石、实木地板、黑镜、灰镜
面积：600 m²

Project Name: Guangzhou Xinghe Dandi Villa
Design Company: Matrix Interior Design
Materials: Volakas Marble, Peacok Golden Marble, Solid Wood Flooring, Black Mirror, Grey Mirror
Area: 600 m²

本案为一个别墅样板房，总共三层空间。整体风格既现代又极富东方气质，在材料色彩的运用上，以灰色为主色调，如灰色石材地面、灰色的墙面布艺，并加饰一些艳丽的花卉，使现代简洁的空间多了一丝温馨与活泼。作为看够了米黄色系的客户来说，会有一股清新自然的感觉。值得一提的是别墅的地下层，山地形态的地下室层高非常高，采光和通风极好，因此也为整个别墅增添了不少的精彩之处。

地下层平面图 / Basement Plan

一层平面图 / First Floor Plan

The design is for a model villa of three storys. It's modern but oriental. The main color is grey, grey marble floor and grey fabric wall. Yet a few bright flowers added to the grey space give it feelings warm and lively. For customers tired of beige, this is an eye-refreshing space. It is worth mentioning that the basement is of an impressive height with great lighting and smooth ventilation, which makes the design another finishing point.

二层平面图 / Second Floor Plan

三层平面图 / Third Floor Plan

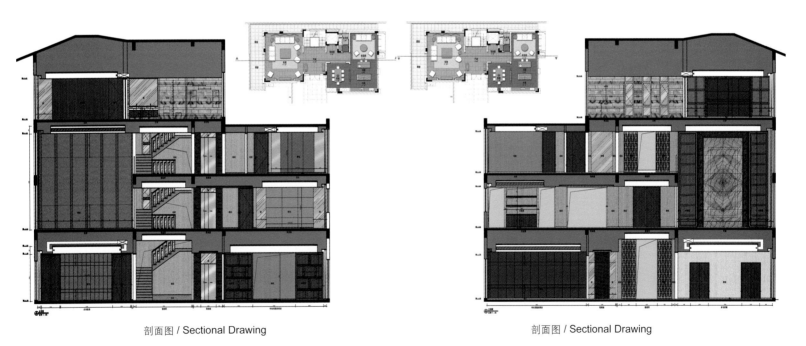

剖面图 / Sectional Drawing   剖面图 / Sectional Drawing

谁解茶中味，水墨也生香
Who Appreciates Tea's Taste? Chinese Ink Painting Generates Fragrance

项目名称：深圳招华曦城六期叠加别墅样板房 KENZO 之家
设计公司：戴勇室内设计师事务所
软装设计：戴勇室内设计师事务所 & 深圳市卡萨艺术品有限公司
用材：金镶玉云石、雅士白云石、西班牙仿石砖、镜面黑色不锈钢、麻草壁纸、铁刀木、真丝布艺、仿古橡木地板、KENZO 品牌家具
面积：350 m²

Project Name: KENZO Home
Design Company: Eric Tai Design Co., Ltd.
Upholstering Design: Eric Tai Design Co., Ltd. & Shenzhen Katha Artwork Co., Ltd.
Materials: Marble, Antique Brick, Stainless Steel, Fibre Wallpaper, Kassod Free, Silk, Oak Flooring, KENZO Furniture
Area: 350 m²

东方风格一直是表达不完的主题，在本案设计中设计师融入了更多的当代设计元素及抽象图形，表达当代东方的空间意蕴。

功能性的最大化是设计之初的首要需求，本案设计了客餐厅、品茶室、中西厨房、家庭厅、书房、藏衣间、影视厅、酒吧区及四间卧房。卧室的坡屋顶挑高也是空间的一大亮点，通过天花的造型处理营造出浪漫的度假感受。三处屋顶天台更是让主人可以随时随地享受灿烂的阳光好去处。

不同材料的运用让空间呈现出不同的视觉效果，极大地丰富了东方风格设计的手法及工艺，让空间体现出新的视觉形象。

水墨纹的仿石砖、黑白的主体色彩、中国红元素、紫铜的莲花挂饰、围屋状的品茶室、金属假山石雕塑、年轮地毯、当代水墨、新颖的 KENZO 品牌家具及富有异域奢华感觉的床品等，诸多设计元素最终融为一体，统一在一个既东方又当代的黑白灰的尊贵氛围中，间或跃动的一抹红色、橘色或黄色，带给人不间断的愉悦感受。

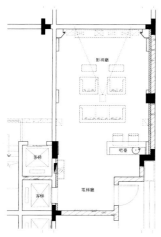

地下层平面图 / Basement Plan

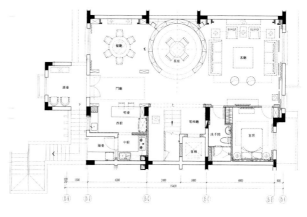

一层平面图 / First Floor Plan

二层平面图 / Second Floor Plan

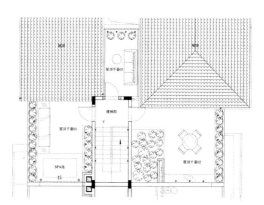

屋顶平面图 / Roof Plan

As Chinese style has always been a theme that cannot be used to its full, more elements of design and abstract pattern are aimed for creating charming eastern space.

Maximized functions are the primary need to be solved, which have been realized very well by arranging the living and the dining rooms, tea rooms, western and eastern kitchen, family hall, study, cloak room, visual room, bar and four bedrooms. The raised sloping roof makes another light spot, where the ceiling makes romantic feelings of vacation. On other 3 rooftops, the family can enjoy sunlight as long as they want.

More materials are used to allow for more visual effect while enriching the approaches and process of the eastern style design, the space thus having a new visual image.

Ashlar tile of Chinese ink painting, black-white hue, element of Chinese red, lotus hanging drop of red copper, tea room like a embracing house, and sculpture of metal rockery, are all fused into a unity in an eastern setting of black, white and grey. But then, colors of red, orange and yellow exert impacting surprises one after another.

璧月琼枝梦秦淮
On Night When the Moon Waves Tree Branches, Qin-Huai River Is Dreamed.

项目名称：无锡复地样板房
设计公司：矩阵纵横
用材：黑金沙、白色皮革、马赛克、手绘墙纸、银箔饰面
面积：120 m²

Project Name: Wuxi Forte Show Flat
Design Company: Matrix Interior Design
Materials: Black Galaxy Marble, White Leather, Mosaic, Hand-Painted Wallpaper, Silver Foil Veneer
Area: 120 m²

将古韵风加入这套靠近京杭大运河的平层住宅中，无锡本地的秦淮风景与当下生活便悄然融合。整体空间定位为新中式，以染色柚木、墙布、白洞石等材料来营造内敛的空间氛围；素雅、温馨的深色调，让中式韵味在空气中弥漫，沉稳中透露出现代的生活气息。

Fused into the bungalow space by the Grand Cannel is the archaic Chinese style, as well as the Qin-Huai scenery of Wuxi and the style of the present life. In a setting positioned as neo-Chinese style, materials of dyed tea, textile wall covering and white travertine are employed to create a reserved atmosphere; elegant, warm dark hues permeate everywhere, overspending the Chinese flavor in a steady and modern life style.

靛蓝与明黄铺陈的色彩华章
Graceful Poem by Indigo and Brilliant Yellow

项目名称：惠州央筑花园中式风格样板房
设计公司：KSL 设计事务所
设计师：林冠成
用材：水曲柳、黑钢、草编墙纸、夹丝玻璃
地点：中国广东
面积：142 m²

Project Name: Huizhou Yangzhu Garden Chinese-Styled Show Flat
Design Company: KSL Design Company
Designer: Lin Guancheng
Materials: Ashtree, Stainless Steel, Grass-Woven Wallpaper, Wire Glass
Location: Guangdong, China
Area: 142 m²

端庄清雅、气韵丰华、精致含蓄的东方传统审美境界在本案设计中深沉而优雅的徐徐展开。靛蓝与明黄碰撞出深藏的秀美与雅致；草编墙纸与木纹地板的稳健搭配出醉人的古典气质。整个空间无需过多修饰，雕花的瓦当、华美的丝绸、陶瓷的鼓凳、圆润的瓷瓶，看似随意的点缀，却尽显古雅中式的内敛气韵，既不张扬而又深入心扉。沉静而优雅、低调而韵律的室内设计让人醉心沉迷。

Traditional aesthetics of delicate and refined grace and elegance is carried out in this space. In collision, colors of indigo and brilliant yellow create delicacy and elegance. The staidness of grass-woven and wood grain flooring bring forward intoxicating classics. Free of decorative redundancy, carved tile, gorgeous silk, porcelain stool, and vase are seemingly put, but all are aimed to set off a reserved and conservative Chinese style, secretly going into the depth of heart. The interior design, calm and graceful, low-key and dynamic, makes people lost in its attraction.

悠游写意，水墨人生
Freehand Chinese Ink Painting

项目名称：皇都花园别墅
设计公司：香港 W+S 世尊设计集团
面积：500 m²

Name of Project: Imperial Garden Villa, Shang Hai
Design Company: W+S Design Group
Area: 500 m²

Ben 认为，家的设计首先要营造出一种幸福感。作为设计师自己的家，选择任何一种固有的风格去装饰都不合适，所以 Ben 选择了一个"去风格化"的设计思路。

这所房子原建筑是一个简约通透的"玻璃盒子"，设计师只合理地规划了空间，尽可能地保留了建筑本身的空灵之美。作为一个设计师和中国水墨画家，Ben 喜欢收藏各类艺术品，同时认为家就是一个舞台、一个背景、一个容器，而其中人就是舞者，就是主角，这个容器可以承载主人的成长，所以 Ben 以"GALLERY——画廊"为设计主题，随着主人年龄与阅历的增长，心境的变化，主人可以不断更新室内家具陈设品和艺术藏品。这样，家便可以与主人一起成长。

Ben believes creating family happiness is the priority when it comes to house design.

To design his own house, Ben prefers not to use certain style to frame his design. The original building of the house is a glass box, simple and crystal, Ben kept these two features while he designed the rooms. As a designer and a Chinese inking painter, Ben loves collecting all kinds of artworks, which inspires him and makes him believe interior design is like a stage, a setting and a time machine, and human being is like the dancer and the leading actor dancing on this stage while being recorded by the time machine. That's why Ben takes "Gallery" as the main theme to work on. He hopes with increasing life experience, interior design is able to be in tune with the trend as well as be updated with different art collections and new arrangement of furnitures, by this way, a house can be really alive, lives up to its owner.

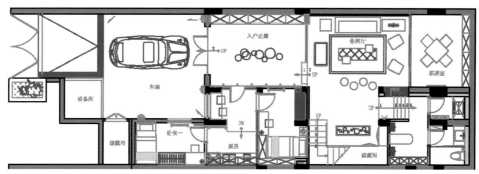

地下层平面图 / Basement Plan

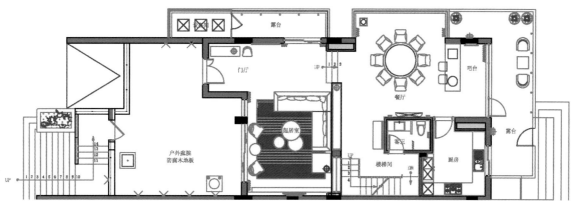

一层平面图 / First Floor Plan

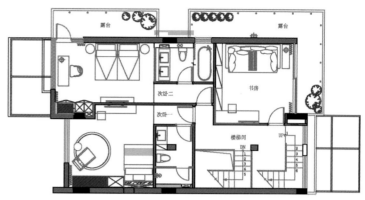

二层平面图 / Second Floor Plan

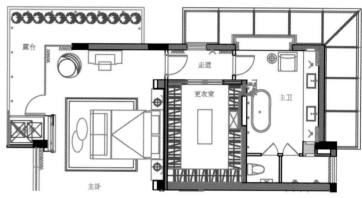

三层平面图 / Third Floor Plan

桃花源里藏芳华，
美在深闺惹人醉

Springtime Is Hidden in the
Peach Garden, Beauty is
Intoxicating in the Boudoir

项目名称：杭州桃花源
设计师：邱德光

Project Name: The Peach Garden, Hangzhou
Designer: Qiu Deguang

新装饰主义大师邱德光在杭州桃花源的设计中，依托中式园林的框架，让中国精神与现代巴洛克、Art Deco 完美对接，使新装饰主义风格成功融入江南园林。面对仿古的园林背景，邱德光没有选择中国古典家具，也没有采用原汁原味的传统语汇，而是与现代生活对接，从各种风格中筛选新装饰主义语汇，让杭州的人文气息与巴洛克、Art Deco 混搭，创造了一个既具有民族个性又兼容现代功能的空间，它以东方元素为语汇诠释着国际现代主义设计风格。

走进杭州桃花源，映入眼帘的是中国园林精华：蜿蜒的回廊、小桥流水、山水洞天、亭台楼阁。深入其中，室内空间的与众不同尽收眼底。西方现代主义的硬件设施，精雕细琢简约图腾的精神内在，西式壁炉对照明式圈椅……这是一种东西混搭的祥和境界。设计师将屋主定位为极具眼光的收藏家，他收藏着中

式园林的精髓，也收藏了具有国际观的完好现代生活。

杭州桃花源挑高的厅堂，将中国文化与现代时尚融合，既有典雅的明式官帽椅与案几，也有 FENDI、Cassina 等来自达芬奇家居的华丽西式家具，在黑金锋大理石、白色洞石搭构的背景基调中，展现一种时代感与明快感，黑白基色透过艺术品、建筑材质肌理，还隐隐透露出中国山水画的意境与灵动，在不经意中，也巧妙地把中国式图纹窗框、地毯等融入进去而不显突兀，反而更显出当代中国雅宅的气度与典范。

在风格独具的餐厅，则兼融大户宴客排场与中国文人精神，糅合出平衡典雅的氛围，璀璨的水晶灯以中式鼎的造型打造，整面墙以中国语汇浮雕装饰。同样的精神气度也贯穿于每个空间，沉稳气韵的书房有着宫灯造型的天花灯具、时尚复古的书桌椅；温馨秀雅的起居室，陈设着具有东方灵感的当代变异水墨画、图绘花卉图案的 KENZO 单椅；明丽卓然的主卧，有着典型 Art Deco 图纹壁面，天花并以弧形打造包厢感，把 Art Deco 与新东方精神发扬到了极致。

游走于古典文气与现代时尚之间，在东方的温雅上增加西方的生动，在西方的明丽之上添入东方的气韵，一切的一切都在努力营造东西兼容并蓄的独特魅力与气质，让具有国际视野与胸襟的现代中国人得以抒怀。

By Qiu Deguang, a neo Art Deco great master, the interior design for the Peach Garden, Hangzhou, perfectly pieces China's spirit and Art Deco in a frame of Chinese garden. This successfully implants new Art Deco into gardens of low reaches of Yangtze River.

In the face of an antique garden background, Qiu Deguang makes a butt joint with modern life and screen of new Deco vocabulary from a variety of styles instead of employing Chinese classical furniture, or using authentic traditional vocabulary, so that local cultural atmosphere, Baroque and Art Deco are mixed to create a space with personality, compatible with modern function. An interior it makes that belongs to the world and is originated in Eastern inspiration and interprets international modernist design.

The space features essence reproduction of Chinese garden: winding corridor, bridge, landscape, cave and pavilions comes into view. Then comes a unique interior where to accommodate western modernism hardware, crafted totem, Western-styled

fireplace and lighting chair. A peaceful realm the combination of the east and the west accomplishes. The homeowner is supposed to be a collector with a world-wide view, who collects here Chinese gardens and modern life with international perspective.

The high-ceilinged hall, specifically integrates Chinese culture and modern fashion. Chinese items like arm chair and long tables of Ming style and ornate Western-style furnishings like FENDI and Cassina, indicate modernity and a refreshing pleasure in a setting of black marble and white travertine. Black and white colors take art pieces and architecture material texture as carrier to faintly reveal the ethereal of Chinese landscape painting. Inadvertently yet cleverly, Chinese-style pattern frame and carpets are integrated, as expected to show the elegance and tolerance of a contemporary Chinese house.

The unique restaurant enjoys a banquet field of an old-renowned family and the spirit of Chines literati, creating a balanced but gorgeous atmosphere. The crystal lamp is endowed with a form of an ancient Chinese vessel. The wall is decorated with reliefs of Chinese vocabulary. In the staid study that has retro and fashion desk and chairs, the ceiling is fixed with lighting shaped into palace lantern. The living room is beautiful and refined, and furnished with variations of contemporary Chinese ink painting which is oriental-inspired, and floral KENZO chair; the master bedroom is embellished with wallpaper typical of Art Deco; the curved ceiling creates a sense of boxes. Both promote Art Deco and new Oriental spirit into a new realm.

Between the classical and the modern fashion, the genteel of the east is added with the vivid of the west, and the bright of the west is inserted with the artistic conception of the east. All are sparing no effort to create an inclusive temperament, so that Chinese with global view and open mind can pour out their heart here.

冷色系打造，时尚花开
Cool Color Creates Fashion Prosperous

项目名称：圣莫丽斯 C11 栋 22A 样板房
设计公司：矩阵纵横
设计师：王冠、王志寒
参与设计：王勇、麦海龙
软装执行：周晓云、邓万里
用材：大花白、银河星钻、新西兰灰、实木地板、黑镜、皮革
面积：350 m²

Project Name: St. Mollis 22A Show Flat
Design Company: Matrix Interior Design
Designer: Wang Guan, Wang Zhihan
Participant: Wang Yong, Mai Hailong
Upholstering Design: Zhou Xiaoyun, Deng Wanli
Materials: Marble, Solid Wood Flooring, Black Mirror, Leather
Area: 350 m²

圣莫丽斯定位为深圳中央别墅区高端豪宅，位于塘朗山生态风景区内，塘朗山属永久性、不可再生的自然生态资源，且离福田中心区直线距离约5千米，是真正的"离尘不离城"，进可享福田中心区、香蜜湖、华侨城的繁华，退可拥自然原生态的宁静生活。

由矩阵纵横担纲设计的 C11 栋 22A 是圣莫丽斯的示范单位，其空间塑造轻妆淡抹、不讲究艳丽、浓重的色彩晕染，反以冷色系突出较强的空间感。黑、白、灰主色调搭配亮光的材料，既不会显得过分奢华，又让人感觉很有档次，是对精致优雅的时尚情怀以及工艺雕刻的细腻精神的深度挖掘。

公共空间的设计褪去了缤纷的色彩，如出水芙蓉一般，将简约与卓尔不凡的高雅气质化为一种低调的奢华，冷静而富有至酷魅力，给人留下深刻的印象，使

家居历久弥新。家具配置简约却不单调，独特的光泽使其更显时尚，给人一种舒适与美的享受。在配饰上，设计延续了整体风格，以简洁的造型、完美的细节，营造出时尚、前卫的感觉。

同样，卧室亦未营造出亮丽的视觉效果。温暖的大地色，温文尔雅，给人一种宁静舒适的感觉，留给屋主一个安静的思考空间。

A project this space is located in a high-end community in central Shenzhen, an ecological beauty spot that can't be developed permanently. The geographical place with a straight-line distance of 5 kilometers makes a really place that's kept away from urban hustling and bustling while enjoying city convenience around in a peaceful and harmonious setting.

Design for this show flat wears a simple makeup, neither gorgeous nor rendered with heavy colors. Instead, cool colors are used to set off the strong sense. Three primary colors equipped with polished materials are not over luxurious but appear with more levels. That reflects delicate and elegant feelings and a depth of fine and smooth exploration into craftsmanship.

The public space looks like lotus rising above, transferring simplicity and nobility into a low-key luxury. So calm and charming is it that it impresses others very well. Furnishings therefore can last refreshing. Though simple but not dull, furniture with its unique glister that compliments fashion, offers comfort and aesthetics. Accessories continue the whole style, concise modeling and perfect detail to bring out sense of pioneering fashion.

Similarly, the bedroom is endowed with peace and quiet. The tranquility of earthy hue, gently and cultivated, completes the occupants space for thinking.

简约苏州印象，穿越文人意境
To Simplify Impression of Suzhou, to Traverse Prospect of Literati

项目名称：苏州仁恒·棠北"天涟"墅
设计师：邱德光
参与设计：刘家麟、陈惠君、骆盈桦、廖佩晶、刘永懋
摄影：T-MEDIA
用材：莱姆石、玉石、云石、黑檀木皮、橡木皮、皮革、镀钛板、壁
面积：1 060 m²

Project Name: Sky Ripple Villa
Designer: Qiu Deguang
Participant: Liu Jialin, Chen Huijun, Luo Yinghua, Liao Peijing, Liu Yongmao
Photography: T-MEDIA
Materials: Marble, Veneer, Leather, Titanizing Board, Wallpaper
Area: 1,060 m²

设计大师邱德光最新力作苏州仁恒·棠北"天涟"墅，体现了其"新装饰主义"的理念，把中国园林渗透于空间设计，将当代艺术的简约印象、天然无敌的湖景无缝相接，独创出具有中国文人风格、禅意境界的"当代艺术馆"。

苏州古典园林作为中国园林的代表，被列入世界遗产名录，而中国的造园艺术又与中国的文学和绘画艺术有着深远的历史渊源。苏州仁恒·棠北"天涟"墅的建筑设计将苏州园林的转折、穿透、一步一景、"柳暗花明又一村"的意境在空间中表现出来。设计师在本案设计中，使空间与生活互动，将当代中国文化艺术与外在环境完美对接。

邱德光的"新装饰主义"理念，因应时代的混搭潮流，是一种海纳百川的设计取向。它可以东方也可以西方，可以当代也可以古典，可以华丽也可以极简，所以"天涟"墅的当代极简风格也与众不同，极简中带入中国文化艺术的内涵与精神，而又无一般极简主义的千篇一律、枯燥无聊。

仁恒·棠北一案最大的特色是直面原生湖景——独墅湖，坐拥私家湖岸，邱德光善用此优势，将湖、流水、云等中国文人喜好的意境与道家思想融于一体，让里外天然贯穿、天人合一。

流水、涌泉、禅定、风云、梦云、祥云，更是邱德光与艺术家们在"天涟"墅精心创作的元素，共同组合成强而有力的当代中国艺术。空间及部分进口家具在此俨然已成为配角，默默地衬托出中国人文及艺术的价值，显现当代内敛奢华的气度。

从玄关的开场就可以看出"天涟"墅的独特与匠心别具。水池、穿透式设计、呈现苏州园林的印象与格局 涌泉、艺术家杨柏林创作的飘浮在水上的佛陀雕塑、邱德光设计的"风云——功夫椅"、梅花画作、端景柜，彼此间相呼应、对话，展现出中国文人式的空灵世界，有种"笑看人间我独醒"的气宇，让屋主回到家中，能排除外界之纷扰，平和内心之涌动。

玄关以透明玻璃为饰，尽纳湖光、水色、天光、云影、倒影等多层次的风景，设计师希望借由空间与艺术的碰撞，使不同观赏者有不同的解读，表达"艺术"

的可贵性，展现当代中国艺术的空间美术馆意境。

从玄关走到客厅，经由庭园可走到湖边，只有中国园林格局手法可彰显此特色，一脉相连的客厅、餐厅、吧台，与户外湖景穿透合一，水的意象也被引入客厅，艺术家杨柏林的黑白油画以及墙、地板，都呼应了波光粼粼的水纹印象。玻璃、不锈钢、大理石等材料的运用，使空间宽敞通透，室内石材与建筑外墙一致，达到天人合一的境界。客厅极简风格的家具，陈设皆以无与伦之间的色阶调色，既展现中国文人画的雅致，又不缺当代的时尚感，邱德光的原创椅"梦云"在这里更扮演重要角色，突出的造型与色彩，带入中国文化的温度，画龙点睛地让空间活了起来。

吧台打造宛如云彩的黑白华丽水晶灯，厨房也中西合璧，在中厨与西厨之间，夹着一个可供休憩的空间。茶室的设计更是神来之笔，茶桌以无斧凿修饰的红花梨木，搭配镜钢支撑的桌以及同样镜钢材质的时尚椅，少了传统中国元素的沉重感，保留了自然感，如入当代中国禅之意境。

一楼父母房、二楼主卧，延续中国文人风格概念，低调色系书房，透过邱德光带入中国明式椅概念的"祥云"，更透露中国传统文化质感，主卧一角的休憩空间与外面湖面连成一气，搭配卡希纳品牌日本设计师设计的名椅，鲜绿椅套与湖光呼应。

一楼起居室以多彩油彩画，呼应流水意境。小型多功能厅把麻将桌功能含入，地下室的休闲空间与会客室，更导入活泼时尚氛围，半圆形的玻璃扶手大理石楼梯，宛如当代雕塑创作。楼梯旁设计为传递杯影叠映感的酒窖空间，收藏室两张"梦云"椅点出中国式时尚感。泳池旁打造梦幻的马赛克屏风，把时尚、艺术融为一体。在这里，既时尚，又现代简约，还隐含中国文化、艺术韵味，铺陈出一种当代中国休闲艺廊的新体验。

Can Suzhou Garden be art up or minimal? Sure, it can. And this project, the newest masterpiece by Designer Qiu Deguang makes the concept into reality, where to embody his design philosophy of neo decorationism and seamlessly piece together penetrating design of Suzhou Garden, simple impressionism of modern art and lakescape. This creates a modern art hall with style of Chinese literati and Buddhist mood.

As a representative of Chinese classical garden, Suzhou Garden is inscribed onto the World Heritage List. Chinese garden art and Chinese literature and painting have been connected historically. The interior design makes good reference for turning, demolishing, penetrating as well as approaches like one step makes one scene and there is a way out of where steps come to an end. Mr. Qiu makes a best interpretation for the communication and interaction among contemporary Chinese art and culture, external environment, and life.

Qiu's concept of New Deco follows the mixing trend in the global village era. A deign it is to be tolerant of style diversity. Design can be both eastern and western, either modern or classical. And it can also be gorgeous or minimal. So though minimal, this project is distinctive. The meaning and spiritual implantation of Chinese culture and art makes the space free of baldness and sameness.

Sitting on a private lakeshore, the project features a native lake view. The geographical is made the best use of to introduce lake, water, preferred by Chinese scholars, coupled with Taoist thought. The space thus becomes unified into one from the external to the internal.

Elements of flowing water, spring, meditation, flowing cloud, dreaming cloud and cloud auspicious are carefully created for presenting a powerful contemporary Chinese art gallery. Space and some imported furniture has now shrined into a supporting position, quietly bringing out the value of the Chinese humanities and arts to show restraint and tolerance of contemporary luxury.

The entrance makes a prelude of the spatial ingenuity where pool of transmission exposes impression and pattern of Suzhou Garden. Springs, Buddha sculpture floating above water by artist Yang Bailin, Kung Fu chair by Qiu Deguang, plum painting, and side view counter, are echoing with each other, showing an ethereal world of Chinese literati style. This implies that in the chaotic world, only I can keep sober. The occupants can get into a concentrated and static state.

Through the transparent glass, lake, water, sky, cloud, and reflection, comes into the entrance resulting in scenery of more levels. The collision of space and art is aimed to have different interpretations for different viewers to express the precious nature of art and show a gallery mood of contemporary Chinese art.

The entrance leads directly into the living room. The garden takes access to the lake. Only techniques of Chinese garden can highlight such a feature. Though individual and independent, the living room, the dining room, the bar, and the lake view are unified into one. Imagery of water is introduced into the living room, where black-white oil paintings by artist Yang Bailin, walls, and floors are echoing with the sparkling waterline. Natural elements of glass, stainless steel, and marble are consistent with the façade. This makes a realm that man is an internal part of nature. In the living room, the minimalist furniture and furnishings are coated in black and white, showing the elegant of painting by Chinese literati. Dreaming Cloud Chair by Qiu Deguang plays an important role in highlighting the color of shape, bringing in colors of Chinese culture. And the space comes vivid like a painted dragon comes alive with eyes.

The bar is designed like a black-white crystal lamp. The kitchen is eastern and western, between the both of which is a sitting room. The tea room makes a delicate touch. Tea table is of rosewood and without any traces by pitching tool. The steel-mirror supporting table accompanied with fashion chair of the same steel, keeps away the stiffness of traditional Chinese elements, but preserves the natural sense, making here a contemporary Zen of China.

The room for the parents on the first floor, and the master bedroom on the second floor complete continuation of Chinese literati style. In the low-key study, Ming-style chair of Cloud Auspicious by Qiu Deguang reveals the texture of traditional Chinese culture. One corner of the master bedroom is linked together with the lake, and Cascina chair with its green color echoes with the lake light.

On the first floor the living room employs more oil paintings to echo with the mood of water. The small multi-purpose hall offers mahjong table. The recreation space and the meeting room in the basement take in stylish and lively atmosphere. The semicircular glass handrail and the marble staircase

are like contemporary sculpture. By the stairs next is a cellar. The boxroom has two dreaming cloud chair to point out the Chinese-styled fashion. The pool offers a fantastic mosaic screen to integrate fashion and art. Here is stylish, modern and yet minimal, implying Chinese culture and art taste to lay out a new experience of casual contemporary Chinese gallery.

鲜嫩欲滴的活力家居，
令人怦然心动

Home Furnishing of Vigor and Vitality Makes People Eager to Accomplish Something with Excitement

项目名称：武汉华润中央公园
设计公司：大武汉东三吉设计、武汉澳华装饰设计工程有限公司
设计师：陈洁
摄影师：章浩
用材：水曲柳饰面板、墙纸、实木地板、定制雕花
面积：95 m²

Project Name: China Resources (Wu Han) Central Park
Design Company: Wuhan Dong Sanji Design, Wuhan Aohua Decoration Engineering Co., Ltd.
Designer: Chen Jie
Photographer: Zhang Hao
Materials: Ashtree Veneer, Wallpaper, Solid Wood Flooring, Custom Carving
Area: 95 m²

在前期沟通时，业主清楚地表达了其要求：喜欢不太沉闷、厚重的中式，不用太多墙纸就能表现色彩的层次感，希望居室能有一种年轻的活力，给人越来越年轻的感觉。因此，在本案设计里，设计师将古典的情怀与现代时尚相结合，展现不一样的家居格调。艳丽的色彩、简单的造型，完美体现现代新中式风格。

玄关作为一个缓冲过渡的地段，是客人从繁杂的外界进入一个家庭的最初感觉。可以说，玄关设计是家居装饰设计开端的缩影。本案玄关除了基本的使用功能和储藏功能外，更有风水上的讲究。造型葫芦，象征着招财进宝；红色，开门见红，寓意喜庆和吉祥；瓶子，象征着平平安安；两边的麒麟，可以驱邪镇宅；顶上的吊顶，象征着天圆地方，

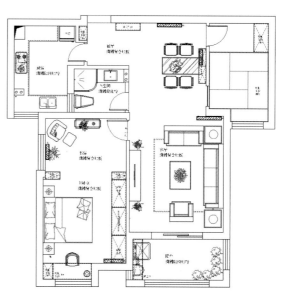

平面图 / Site Plan

云腾吉祥。

客厅别有一番现代韵味，湖蓝色和红色的家具，白色的雕花装饰，浅色的布艺沙发配置多彩的抱枕，红色的电视墙，都显示出中式特色，不仅造价不高，而且能突显居者的年轻心态。餐厅经典的蓝白色对比，彰显出空间时尚美感。中国娃娃的挂画，融喜庆而又浓郁的中式元素于内，呼应空间的主体色。餐厅的吊灯、色彩和造型也都显得格外时尚、美观。

卧室设计别具一格，整个空间以红白色为主色调，采用蓝色作为空间的点缀色彩，让中式家具的古典气质，又有了另一种演绎。被单以及门帘小碎花的图案、原木色的书桌以及中式的吊灯，同样的风格与设计，令空间整体非常舒适。榻榻米的设计，实现了一屋多用的功能，是休闲的好去处，又可以当做客卧，接待客人。床头柜上的莲花台灯更为整个房间增添了清新、和谐的气氛。

本案主体色的把握，背景色的穿插，颜色明度的统一，色彩的冲撞，让整个家居的气氛鲜嫩诱人。

With personal preference for Chinese style, heavy but not too stiff, of vigor and vitality, feelings of being younger and younger instead of employment of too much wallpaper offer a combination of the classical and the modern in a unique tone. Gay colors and simple modeling perfectly embodies new modern Chinese style.

The foyer, as a transitional space, exposes guests into a dwelling space free of bustling and hustling, making a prelude to start the interior design. Besides basic functions like using and storing, the foyer stresses more on Feng Shui. The calabash symbolizes wealth and treasure coming. The red means that happiness can be met with door opened. That is joyous and auspicious. Bottle, stands for safe and peace because of its Chinese pronunciation of Ping Zi, similar to safe's sound of Ping'an in Chinese language. Kylin statures on both sides drive out evil spirit and protect home. And the suspended ceiling makes an emblem that the heaven is square and the land is circle, and that the auspicious rises above cloud.

The living room is modern but unique. Furnishings of light blue and red, decorative carving of white, fabric sofa of light color and colored cushions, and red TV backdrop are more Chinese. It's medium-priced and more reminiscent of young mentality. The sharp contrast of classical blue and white sets off aesthetics that's fashion and stands careful reading and appreciation. Wall paining of Chinese doll is jubilant, echoing with the dominating color within. The droplight is equally modern and good-looking in both terms of color and modeling.

The bedroom is dominated with red and white while blue appears now and then, classical furnishing of Chinese style is interpreted different. Pattern of bed sheets and shivering door curtain, wood desk and Chinese droplight continue the same style and design, so the space is wholly snug as a bug in a rug. The tatami allows for more function in a single room, making an ideal place for leisure and guest accommodation. Table lamp of lotus flower enhance the freshness and the harmony.

The accurate grasp of the main color and the backdrop hue accomplish a lightness unity, where the colorific collision brings forward a refreshing vigor and vitality for the home furnishing.

风华内敛，气度雍容
Reserved and Graceful

项目名称：万科悦湾 A2 复式洋房
设计公司：矩阵纵横
用材：霸王花大理石、麻石、黑钢氟碳漆、绒布硬包、藤编、椰壳板
面积：350 m²

Project Name: Vanke (Pleasure Bay) A2 Duplex House
Design Company: Matrix Interior Design
Materials: Marble, Granite, Black Steel Fluorocarbon Paint, Lint, Rattan Weaving, Coconut Board
Area: 350 m²

"新亚洲"的风格定位，"复古"、"奢华"在这个作品中得到很好的体现。在大面积灰色调的烘托下，深色木饰面与细节处金漆的勾勒，实木断面与绒布硬包的材质对比，突出了空间与众不同的气质。陈设软装在整个空间中起到了画龙点睛的作用。

A project this space is where the position neo-Asian, retro and luxurious is fully presented. With an atmosphere of grey, the delicate dark wood veneer and subtle golden paint, as well as the contrast between solid wood section and decorative velvet reflects a unique taste. And the upholstering makes a finishing point for the space.

亲近自然，轻松怡然
Get Close to Nature, Easy and Happy

项目名称：香港皇璧
设计公司：郑炳坤室内设计
设计师：郑炳坤
面积：310 m²

Project Name: Hong Kong Westminster Terrace
Design Company: Danny Cheng Interiors Ltd.
Designer: Danny Cheng
Area: 310 m²

本案为复式住宅单位，设计师以精致的设计手法突显属于本案的独特感觉。一道电动大门带你走进这个气派的住宅，大门旁的镜面墙身在可增加空间感之余，亦展现了饭厅的景象，增添了趣味。开放式的厨房配以设有洗手盆的长形白色餐桌，带出开放互动的感觉。木皮条子旋转门成为客厅的焦点，可为屋主和客人将客厅打造成一个隐私度较高的空间。全高镜钢电视柜可让屋主在摆放东西之余，亦提升客厅的格调，突显楼底高的优势。设计师以云石作客厅、饭厅的地台，并将其伸延至户外露台，营造高雅时尚的感觉。客厅的花形图案地毯呼应露台的植物、流水声，增添了大自然气息与写意的气氛。由厨房、饭厅到客厅，没有间隔墙身，整个空间富有通透感及连贯性。

楼梯及睡房铺设了木地板，为休息的空间营造出温暖感。主人房以棕色作主色调，带出和谐的气氛。床背以扣布作墙身，给人以舒适的感觉。偏厅提供充足的空间给屋主休息及放松心情。花形图案地毯为主人房增添了生气，并与客厅互相呼应。以木皮作主要物料的衣帽间，充分地利用空间，让屋主可摆放大量衣物，配合玻璃面的独立柜，可陈列首饰、手表等饰物，美观与实用并重。

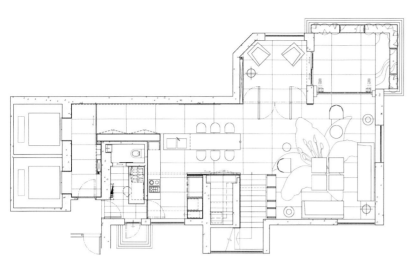
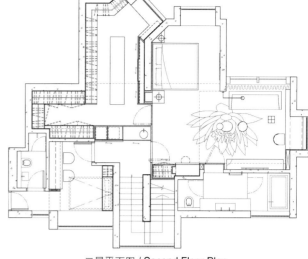

一层平面图 / First Floor Plan

二层平面图 / Second Floor Plan

A duplex dwelling place this project is with an exquisite design highlighting unique feelings.

After an electric gate is a stylish space. That is where this project is. Wall mirror by the gate door increases the spatial sense, also reflecting the dining scene and promoting fun and interest. The open kitchen has an elongated white dining table fixed with a wash basin, which brings out open and interactive feelings. The revolving door of veneer strips has become the focus of the living room, creating a higher degree of privacy for the owner or guest. Full-height TV cabinet of mirror steel makes the best use of the height, not only for storage, but also complimenting the living style. The living room is endowed with a marble platform, which extends onto the terrace. That builds up fashion and elegance. The floral carpet in the living room echoes with plants and sounding water on the terrace, adding natural flavor and impressionistic atmosphere. Among the kitchen, the dining room and the living room, no wall serves as partition, the entire space thus being full of a sense of permeability and continuity.

The stairs and the sleeping floor are fixed with wooden flooring, completing a sense of warmth for the resting area. The master bedroom is coated in brown, bringing forward a harmonious atmosphere. The backdrop for the bed is wrapped in palpable cloth, complimentary to comfort. The side hall provides sufficient space for the owner to rest and relax. Flower-patterned carpet adds vigor and vitality to the main room in echoing with the living room. The cloakroom mainly uses veneer. Here presents a room that makes the best use of the space with jewelry and watches displayed in independent cabinets. That's both beautiful and practical.

岭南风情韵悠悠
Amorous Feelings Come Long-Drawn-Out from the South of Five Ridges

项目名称：佛山龙泉豪苑 11#2101 户型
设计公司：广州市韦格斯扬设计有限公司
面积：110 m²

Project Name: Dragon Spring Mansion, Foshan
Design Company: GrandGhostCanyon Designers Associates Ltd.
Area: 110 m²

进入大门，映入眼帘的是玄关，以传统中式窗花元素提炼成的屏风造型吸引视线。屏风背面暗藏光源，为整个玄关增添光影气氛，并使屏风前陈设的艺术品更显时尚。鞋室门采用折叠门的形式，不仅能将玄关整体修正得平整统一，还能将岭南传统生活气息透露出来。

本案的客厅拥有一般豪宅都没有的 3.4 米层高，整体高挑开阔。客厅天花由一条光带与一条黑镜钢勾画而成，简约大气，与国外进口的庞贝啡原石地花相辅相成。背幅屏风造型延续了玄关的屏风元素，并且拼接成折叠形式，有背靠屏风而坐的帝王居室之气势。屏风造型底面采用粗犷的艺术肌理漆，并用暗藏的洗墙射灯照射出来，凸显细节设计，并为客厅整体灯光效果增添了一种淡雅气氛。穿过客厅，来到该户型景观最好的阳台，这里被设计师打造成一个喝茶、赏景的休闲区域，该区域三边浅水环绕，水池底设置有水灯，并配以休闲的户外沙发，住户能在一天的辛苦工作之后在这里放松身心地赏景、赏月、看星，也为整个空间平添一种轻松的"禅"意。客厅下方的餐厅，可放置一张八人大餐桌，餐桌上方配水晶梅花管吊饰，奢华大气，并在餐桌中轴线上的下方设置一个艺术雕塑，引客入室，增加住户用餐时的专属感。

本案主卧室是一个带书房的大套间，采用半通透的吊趟门将书房与睡眠区隔开。书房的书架借鉴了传统中式住宅里的博古架造型，利用现代时尚的设计手法表现出来，可放置艺术收藏品、书籍。主卧室背幅以皮革硬包与布艺硬包组合而成。床中黑色真皮硬包，为睡眠空间增添一种宁静气氛。两边对称的硬包造型为主卧增添一种和谐的温馨感。主卫干湿分配合理，厕浴分离，便于日后清洁使用。洗手台采用进口古堡灰石材，墙面用进口英伦玉石点缀，高贵大气，彰显住户时尚、奢华的品位。

平面图 / Site Plan

The entrance features a screen of traditional Chinese grille. So eye-catching it is. Behind the screen is hidden light source, adding lighting atmosphere for the entire entrance. The lighting, on the contrary, compliments the fashion art of the art pieces before the screen. The shoe compartment is fixed with folding door, not only making the entrance more unified, but also revealing the local traditional life.

The living room enjoys a rare height of 3.4 meters. The ceiling consists of a band of light and a black mirror steel, minimalist and grand to be complimentary with the external flooring of imported marble. The screen continues the style of that one in the entrance, and is spliced into a folded form. That makes an image of an imperial bedroom, sitting against the wall. The screen underside employs rough art paint, irradiated with hidden spotlights, highlighting the details and adding an elegant atmosphere for lighting effects.

The living room is attached to a balcony with the best view, a place transferred into a casual family room for tea drinking and scenery enjoying. The family room is surrounded with a shallow pool on three sides. Water lights are fixed onto the pool bottom. The sofa ensures occupants soul relaxation after a day to appreciate the scenery, the moon, and the stars. This makes Zen for the entire space. Beneath the living room is a dining table large enough for 8 people. Above the table are crystal plum pendants, luxurious and grand. Under the place along the central axis of the table, there is an art sculpture, which increases an exclusive dining experience.

The master bedroom is a large suite with a study, both of which are partitioned with a semi-transparent door. The book shelf in the study makes good reference to antique-to-curio shelf traditionally used in Chinese house but now presented with modern and stylish design techniques, where to store art collections and books. The backdrop for the master bedroom is made of leather and fabric upholstering. The leather decoration for the bed adds peace and quiet for the sleeping. The symmetrical shapes add a harmonious sense of warmth. The dry area and the wet area of the bathroom is kept independent. Washbasin is of imported grey stone while walls are decorated with marble from British, elegant and grand, highlighting the sense of luxury and fashion of the space.

凭水临风逍遥居
Fancinating Residence

项目名称：比佛利山庄
设计公司：博斯韦尔建工

Project Name: Beverly Hills Residence
Design Company: Boswell Construction

比佛利山庄东北角有一个斜坡，斜坡之上，有一栋建于20世纪中期的建筑。借助于洛杉矶本土建筑公司"博斯韦尔建工"的妙手，曾经的沧桑，如今已华丽丽地加冕成现代的优雅之所。

虽然只是其中之一，但是流线型的直线建筑形式、中性的色调不禁让人回想起中世纪"劳斯代尔庄园"的魅力。大大的圆形开窗，曾经的痕迹，隐现于建筑鲜明的白色外观之上。

穿过铁艺的自动大门，绕过LED照明的水池，便进入了本案予人深刻印象的家庭空间。

自恢宏的玄关起，定制的金属涂层屏风，优雅地围成圆形的图案，并以其富有吸引力的雕塑质感回应20世纪中期建筑的宏伟框架，也再现当时的城市风光。宽大的房间横向布局。主卧套房、私人卧室区与开放式的厨房、餐厅分居两侧。后面抬高的空间，3米高的玻璃折叠门遁形于连墙之内，完美地衔接着室内外空间。

主卫，配有独立的浴缸，其地面采用带有精美纹理的品牌大理石。蒸汽浴室配有品牌的灯具。两墙之间以滑动的玻璃门相连，门外是壮丽的城市景观。其中一个玻璃门直接通往无边泳池，来往出入变得更为自由、通畅。

在保留庄园旧时"现代主义"的基础上，融入了更为现代的元素。因此，重新装修后的本案，为生活在现代的人们提供了一个充满加州迷人风情的空间。

Set upon a slope in the northeast corner of Beverly Hills sits a mid-century gem that has been entirely transformed to embrace today's standards of elegance by Boswell Construction, a Los Angeles-based company that has constructed many outstanding homes in the area.

The streamlined rectilinear building form and neutral color palette of the residence evoke the original mid-century glamour of Trousdale Estates, the coveted neighborhood of which this home is a part. Only a large circular window, a design element reminiscent of that period, interrupts its stark white facade.

After passing through wrought-iron gates, a concrete motor court and across a shallow rectangular pond subtly illuminated by LEDs, one enters this impressive home.

From the grand entry, custom-designed, powder-coated metal screens in a circular pattern reminiscent of the mid-century period of the building's original architecture frame views of the great room and the sparking city beyond in an appealing, sculptural way.

The impressive great room expands across the linear site, with the master suite and private bedroom areas on one side and the gourmet open kitchen and dining area on the other for a functional flow. 3-meters-tall glass pocket doors stretch across the entire rear elevation and disappear into the side walls, seamlessly connecting the interior with the outdoors.

The master bathroom boasts gorgeous book-matched Calacatta marble, and a freestanding bathtub and steam shower equipped with Gessi fixtures. Two walls fitted with sliding glass doors embrace city views, one of which opens directly onto an infinity pool for effortless access.

This residence respects the integrity of Trousdale Estates' California Modernist origins while adding chic contemporary elements to create a canvas for glamorous California indoor-outdoor living today.

Feel the Eastern Zen Style — Living Space X · 245

隐于山林，归于自然
Reclusive amid Forest and Mountain, Returning to Nature

项目名称：哥伦比亚乡村别墅
设计公司：EnE 建筑工作室
设计师：卡洛斯·努涅斯、纳塔利娅·埃雷迪亚
摄影师：大卫·乌里韦
地点：哥伦比亚比耶塔
面积：650 m²

Project Name: Casa 3 at Colinas de Payande
Design Company: Arquitectura en Estudio with Natalia Heredia
Designer: Carlos Nuñez, Natalia Heredia
Photographer: David Uribe
Location: Villeta, Colombia
Area: 650 m²

本案位于哥伦比亚波哥大以北100千米外的比耶塔小镇。该地天气终年温暖、风景优美、山如眉黛。因此设计有意模糊了室内外之间的界限，将最美的室外景观纳入室内空间。

通过开放的庭院花园，便可到达内部空间。乍一看，整个室内空间宛若隐匿于青翠欲滴的植被中一般。室内客厅、餐厅、厨房全朝向外面的泳池、阳台，似乎整个自然景观延展进了内部空间。

包括4个卧室在内的隐私区，与侧翼相连，通过开放的花园和廊道便可进入。所有浴室尽情沐浴在自然的光线中。包括庭院在内的空间，各卧室全部朝向山野绿丛，真可谓是"风景这边独好"。

圆金属柱、清水平面混凝土板、任由开启的精制木屏，用于本案空间的建筑语言原本就这么简单。家具大多白色砖石结构，便于维护。

山景、绿地、水情无处不在。模糊的内外之间，任由空气自由流通。

100 kilometers north of Bogotá lies the small town of Villeta, where this country house has been built. The ever-present warm weather, the dramatic landscapes and the greenery of the mountains have completely driven the design of the house, which intends to blur the boundaries between the exterior and the interior while framing the most beautiful views.

You access the house via an open courtyard/garden through which you can see, at a glance, the whole span of the social areas of the house while being immediately drawn into the exuberance of the vegetation of the region. The interior living, dining and kitchen areas widely open towards the exterior swimming pool and deck areas, forming a very horizontal, open plan social space that spills into the landscape.

The more private area, which contains the 4 bedrooms, has been laid out in an adjacent wing to be accessed through an open garden/corridor. While all the bathrooms benefit from natural light

and their own private garden, the bedrooms look out onto the mountains from their privileged stance and can be completely open towards the landscape.

The house makes the most out of a very simple constructive language: round metal columns, clean horizontal concrete slabs and a carefully knit timber screen, which can completely open or close at will. Most of the furniture is made up by white painted masonry, simplifying the maintenance process.

The mountains, the greenery and the water are present everywhere, and air is permanently flowing freely from the outside to the inside and from the inside to the outside, although here we can't really know which is which.

平面图 / Site Plan

醉卧花间君莫笑
Don't Laugh at People, if They Lie Drunk in Flowers

项目名称:"花褶边"别墅
设计公司:西仁·帕特尔建筑师事务所

Project Name: The Frill House
Design Company: Hiren Patel Architects

"花褶边"别墅的设计以自然风景作为其重心。专门为冬夏两季设置的不同花园,彰显着空间以"气候"为导向的设计。"Z"字形的平面布局内,南北各有花园。南面花园绿如"森林",在夏季为空间提供了阴凉之所。北部花园,内有大量草坪,充分地利用了从早到晚的日照,为空间提供着阵阵暖阳。

空间每一处都朝向花园景观,不同地方有着不同的花园景观,带给人的却是如诗般的观感。

在玄关阳台,似乎距花园仅几步之远,观感如此之同。不论是往来于走道之间,还是进出于休闲区间,抑或是透过巨大的角窗,皆可见"夏园"的怡然景观。

穿过走道、走进花园,内有玻璃楼阁,即为家庭室。家庭室周围是如棕榈一样高大的热带植物。沿着南墙是巨大、浓密的种植园。

通往"夏园"的是一双高阳台。过了阳台,便是室内空间。再往前走,经过餐厅、书房,便到了空间尽头。尽头处是主卧。卧室东面设有通往"夏园"的阳台,是一个沐浴晨曦、静享下午茶的绝佳之地。

入口车库、客房都可以享受不同的花园景观。空间设计以气候为导向。纯白的基调与浓绿的自然景观相呼应,给人以雕塑般的质感。全屋定制的家具及铺陈有着精湛的技艺。常用的色彩及物件为空间增添了诸多的感官之乐。原创的绘画及艺术品,使空间无形中得以升华。

建筑基本理念与艺术手法的运用及艺术品的陈列赋予了空间"永恒"的品质。

This house is designed with prime focus with landscape. This is a first climate oriented house plan. It has a summer garden and a winter garden. Here the plan forms "Z". Thus it creates two garden spaces on south side and north side, the south side garden is designed like a "forest" to help tree plantations to create shadows on the building. The north side garden has more lawn area, which has shadow of building in afternoon and evening. It gets good morning sun and than later part of a day is shaded.

Each space has a view to garden. And we planned different philosophy to experience the garden from different places. The experience of garden is like poetry. At entry veranda the garden feel is different, where you see garden from arrival room few steps from veranda. Then you enter to passage and out good grader enters inside the house, as one enters informal living and gets a complete view of summer garden from huge corner window. Opposite to that on south side one proceeds through passage and garden on both side, reaches in glass pavilion which is a family room completely surrounded by tall palms tropical plants and huge and dense tree plantation on south wall of the plot.

Then one passes further deep in the house via a grand double height veranda on north opening to summer garden. And further one passed from dining and study to the end of the house which is a master bed room which opens with cozy veranda on east side of the room in summer garden. A perfect place for the morning sun and morning bed tea!

The entry car porch, puja room and guest room also have a different type of garden. The climate directed the plan formations. Out side form is masses of Forms in white. White makes it sculptural quality against the prominent greens of landscape. Interior is well thought of and minutely detailed out with a great workmanship and customized furniture. The Indian colors and objects in interiors bring joy in the house, which breaks soberness. Original paintings and art objects complement the spaces.

Basic principles of Architecture have been applied to the house and use of art & art-objects make the spaces of 'timeless' quality.

水映明月满庭芳

The Moon Is Reflected in Water, the Fragrance Is Stuffing the Patio

项目名称：四合院
设计公司：西仁·帕特尔建筑师事务所

Project Name: The Courtyard House
Design Company: Hiren Patel Architects

本案是典型的院落别墅，另外还附有一个长、宽各2米的网格状空间。设计以"生活起居空间、走道全部朝向开放区域及花园"为主要理念。中央庭院有一泳池，是设计的主要亮点，设计也因此变得内敛、低调。花园虽然位于后部空间，但是无论处于空间哪个位置，视线都可通达。

走过一条3.7米宽、且与中央庭院水池平行而立的走道，便可到达主生活空间与两个卧室。走道尽头是餐厅与起居室。鉴于该区域使用频繁，三面另加了宽大的阳台。阳台的使用，彰显了北侧花园不同寻常的价值。

除了宽大的门窗，本案还设有同样尺寸的天窗，便于引入早晚的光线，尽可能地利用了当地的气候条件。各开口及门窗处设置的屏风、深入的阳台使空间不受到毒热阳光炙烤。傍晚、早上，空气凉爽的时刻，阳台又成了卓越的户外空间。除了碾压混凝土结构的屋顶，阳台、空间上部都覆盖着木板。地面则为天然大理石、木板混合结构。

简单的建筑理念、艺术手法的运用及艺术品的陈列让空间有了"永恒"的品质。

The courtyard house is designed with a grid of 2x2 meters. The house has been designed in such a way hat all the living spaces and passage face open space and garden, which was also the main design concept since its inception, thus creating a central courtyard which holds the reflection pool making it a major design element. This also makes the design an introvert one, leaving garden at backside but still visible and accessible from all the rooms of house.

Formal living room comes first along with 2 bedrooms followed by 3.7 meters wide passage parallel to central courtyard and reflection pool. Passage ends to dining room and family room. This area is maximum used area, thus has a big verandah all three sides, which adds the value to north side garden.

This house has big size of openings and skylight to allow ample amount of light to be entered in house throughout the day. It is also a crucial feature in Ahmedabad climate. Here light is essential but not a scorching heat. Screens and deep verandah protect these huge openings from harsh Sun. Verandahs also become wonderful outdoors spaces during relatively cool evenings and mornings.

Courtyard House is made of RCC roof structure covered with wood. Also verandah and upper rooms are covered with wood. Floors are covered with natural marble stone & wooden planks. Custom designed doors and windows are made of wood and laminated glass.

Basic principles of Architecture have been applied to the house and use of art & art-objects make the spaces of "timeless" quality.

蝶舞香飞水云间
Above Water and over Cloud are Flying Butterfly and Fragrance

项目名称：佛山宏越·云山峰境
设计公司：韦格斯扬建筑
面积：350 m²

Project Name: Cloud Mountain
Design Company: GrandGhostCanyon Designers Associates Ltd.
Area: 350 m²

宏越·云山峰境坐落于南海，位于白云山、象山、龙岗三山及连片水域的地方。自然景观条件优越，配套齐全，交通便利，是丹灶城区最大型的商住项目。本案为4层别墅，为现代东南亚风格的设计。

步入室内，经过宽敞舒适的前厅便可通往餐厅，餐厅的装饰柜造型，有实有虚的处理；客厅为两层打通的空间，给人带来挑高震撼的空间感受，墙身面块的处理，嵌入一些茶色金属的点缀，电视墙则拼接了一幅深色木芭蕉叶浮雕，给整个户型带来厚重沉稳的感受。

负一层为娱乐空间楼层，影视室和酒吧兼备的空间，给业主以不同的生活体验。二楼主要为卧室楼层，设置了起居室空间，配备了一个怡人的内嵌式阳台，提升了整体的档次。三层为主人房楼层，主卧是给人带来深刻印象的房间，天花坡顶的设计，米色墙纸、深色木的对比，营造出素雅稳重的空间。主卧配备了独立的书房空间和衣帽间，让主人分外尊贵。四层设置为品茶室休闲空间，一侧为舒适的沙发组合，另一侧则为高档漆画屏风造型。步入该空间给人以身心舒畅的感觉。

Perched on the seashore of South Sea, this project is located at the conjunctions of three hills of White Cloud, Elephant, and Dragon and adjacent to patches of water. The golden geographical conditions, well-facilitated infrastructures and convenient traffic are complimentary to the only residential space within its neighboring community.

A 4-storey floor villa this space is that is endowed with a Southeastern Asian style.

The large comfort foyer takes direct access to the dining room, where sideboards are treated with transparency-solid approaches. The living room is as high as two floors, waging an imposing visual effect. Walls are embellished

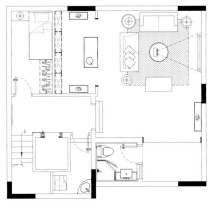

地下层平面图 / Basement Plan

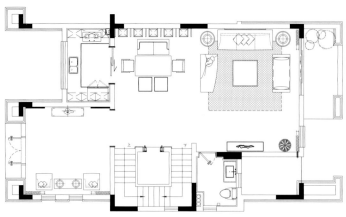

一层平面图 / First Floor Plan

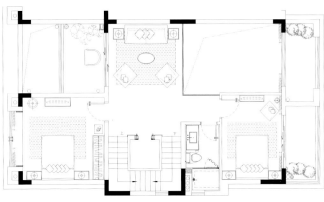

二层平面图 / Second Floor Plan

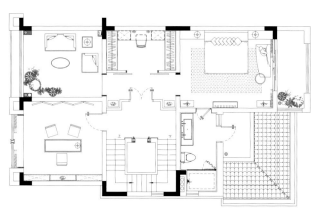

三层平面图 / Third Floor Plan

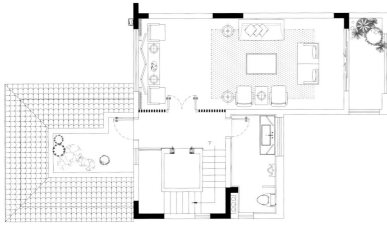

四层平面图 / Fourth Floor Plan

with tawny metals. TV backdrop is collaged with a dark hue relief of banana, offering the space a sense of magnanimity and staidness.

With home theater and bar, the basement is used for entertainment, where occupants can enjoy themselves. The 2nd floor is for bedroom. Besides the family hall, the recessive balcony upgrades the whole space. The 3rd floor accommodates the master suite. Its features are impressive. The sloping roof, the beige wallpaper and the dark hue wood make a sharp but intended contrast. The free-standing study and the cloak room set off the personal nobility of the master. The 4th floor is a tea-tasting space, on both sides are respectively sofa and high-end screen of lacquer painting. There is bound to present comfort to relax body and rejuvenate mind.

酒店，因地制宜
Hotel to Local Conditions

传统与现代融合，开启一场震撼心灵的视觉盛宴

Fusion of Tradition and Modern Starts a Heart-Quake Visual Feast

项目名称：深圳四季酒店
设计公司：HBA设计师事务所、都市实践设计有限公司

Project Name: Four Seasons Hotel, Shenzhen
Design Company: Hirsch Bedner & Associates, Urbanus Design Limited

深圳四季酒店坐落于福田区中心地带，旨在于线条略显硬朗的繁华闹市中，打造一座融现代设计感与自然景致于一体的"都市绿洲"。本案由享誉全球的室内设计公司Hirsch Bedner & Associates（HBA）以及中国最具影响力的建筑师团队都市实践设计有限公司（Urbanus）共同打造，它犹如一栋立于沉静绿洲的高塔，倾泻如柱的自然光，更加渲染出摩登现代的气质。

酒店将现代设计元素融入极具前瞻性结构与设施中，再度诠释深圳作为"联合国教科文组织设计之都"（UNESCO City of Design）的赞誉。从建筑外观上看，酒店打破了传统的直线型结构。它本身是双塔式建筑，却以积木拼搭形态展现在客人眼前。这样的建筑结构除了给酒店客房提供完美的自然采光之外，更创造出一个露天平台，供往来商旅休闲小憩，一览城市风光。

迈过玛瑙石镶嵌的酒店门廊，即是挑高的大堂，宁静舒适之感油然而生。宾客轻踏着编织精美的祥云图案地毯，如有步入仙境之感。祥云元素也融入到酒店内部总体装潢设计与客房内饰之中，令吉祥如意的美好寓意绵延不止。同时，在几何形的建筑中，一朵朵祥云带来亦动亦静的绝妙视觉感受，堪称深圳商业中心又一浓墨重彩的艺术杰作。酒店大堂一头是一件布满整个墙面、令人惊叹的艺术装饰，吸引宾客慢慢走入酒店的艺术长廊。长廊中，一系列精心挑选的瑰丽艺术作品，在考究的布景和灯光中熠熠生辉，为宾客带来绝无仅有的艺术发现之旅。

来到迎宾大堂，白色大理石锻造而成的几何形迎宾台，与其背后的艺术墙呼应，带来静中有动的奇妙感受。大堂中央的水景实现了视觉和听觉的双重享受，为宾客带来宁静平和之感。宾客可直接步入绿意盎然的室外，进入一片舒适怡人的绿洲。透过迎宾大堂的落地窗，露台美景一览无遗，闲坐于此更可任思绪尽情傲游，享受一份难能可贵的悠闲时光。

客房与套房
酒店拥有266间充盈着自然光线的奢华客房和豪华套房，为宾客提供宽敞的起居空间。

客房的设计别出心裁，斑斓的当代艺术装饰与床头精心手绘的祥云图案象征着中国传统文化中的福瑞吉祥，流露出对美好事物的期盼。同时，客房地毯上也点缀着翻卷着的水墨祥云，配以紫色、象牙白和奶白色基调，勾勒出一幅浑然天成的淡雅云图。精挑细选的家具尽显奢华，舒适之余不失质感。

餐饮
馥餐厅
馥餐厅，顾名思义，汇聚了美食的馥郁芬芳，为热爱美食美酒的旅行者提供了全新的去处。餐厅由世界知名的SPIN公司设计，将中国传统与现代设计元素相结合，彰显餐厅质感与精致内装。餐厅入口处的通道由璀璨的地射灯照亮，而天花板上装点有极具艺术感的铁丝质感雕塑。

卓粤轩
作为深圳四季酒店的招牌餐厅，卓粤轩的设计充分考虑到中国人以共进美食来交流感情与升华友谊的重要传统，并对传统用餐礼仪、餐厅内艺术品陈列和酒店总体设计风格加以考量。餐厅门口，顶级彩色骨瓷幻化而成的现代艺术作品，与垂挂于天花板的木质元素相辅互映。柔和的灯光和暖色木质墙壁突显空间私密性。私人包间则被赋予温暖的色调，透过巨大的落地窗，宾客可将令人屏息凝神的全城美景收入眼帘。

逸廊
逸廊坐拥深圳城市的醉人夜景，令宾客尽享游走于现实与虚幻之间的非凡体验。酒吧独特精致的装饰设计，营造出一个远离城市纷扰，释放压力的优雅环境。

宴会及会议场地
酒店设有不同样式与大小的宴会厅，主要分布于酒店3层、7层和29层。其中7层和29层的宴会厅以茉莉、玉兰、牡丹、莲花等花卉为主题命名，尽显中式美感。位于3层的四季宴会厅拥有两间前厅，可根据需求灵活用作私人工作间、用餐室、VIP接待室、婚宴或演职人员的化妆间。所有多功能空间均采用无柱式设计风格，落地窗既能提供明亮的自然光线，更能将城市美景一览无遗。宾客还可从课桌式、回型、U型、剧院式、鸡尾酒会、中式宴会和董事会等会议布置中选择最合适的坐席布局。而7层，面积达150平方米的独特户外露台——悦台可用来举办酒会或婚礼庆典，为宾客留下珍贵难忘的回忆。另外，深圳四季酒店的各个宴会厅可满足不同规模的会议及活动的需求。四季宴会厅最多可容纳360位宾客，是举办商务会议以及婚礼庆典的理想之所，而最小的会议空间则最适合20人左右的小型私密聚会。

康体设施
酒店24小时开放的健身中心为需要倒时差的旅行者提供便利，可随时进行心肺和体重训练，或者在桑拿、蒸汽房或涡旋泳池中放松身心。室内泳池是进行水上运动的最好选择，而较大的室外泳池则是孩子们和家人在水中享受嬉戏游乐时光的好去处。

Located in the heart of Futian District, the easily accessible Four Seasons Hotel, Shenzhen, is aimed to create an urban oasis that fuses the modern design and the natural landscape at the heart of downtown. By HBA, a world-renowned interior design firm, and Urbanus, a China's most influential team of architects, the project looks like a quiet tower rising from an oasis. Its natural light renders more modern temperament.

The combination of modern design elements into the forward-looking features and facilities makes a re-interpretation of Shenzhen as a "UNESCO City of Design". The building exterior has broken away the traditional linear structure. Its original twin-tower appearance has now come into existence in a form of building blocks. Such a structure not only provides perfect natural light for the guest rooms, but creates an open platform for business travelers to enjoy cityscape in leisure.

Through the porch wrapped with agate is the high-ceilinged lobby, where to give birth to quiet and comfortable feelings spontaneously.

Steps on the exquisite carpet patterned with auspicious clouds feel as if it were in wonderland. The overall interior design and the decorating in guest rooms continues the auspicious cloud, so that good luck and best wishes can be carried out. Out of the clouds comes a wonderful visual experience. That makes another great art piece in the commercial center of Shenzhen. At the end of the lobby, there is a stunning art decoration that overspreads the whole area, which attracts guests into the gallery. In the gallery, a series of carefully selected magnificent art works bring visitors a unique artistic discovery in the elegant setting and with lights shining.

In the reception lobby, the geometric reception desk is forged with white

marble, which echoes with the art wall behind, static and dynamic to bring marvelous feelings. The waterscape in the central lobby allows for visual and auditory enjoyment as well as peace and quiet. The lobby interior takes direct access to the greenery outside. A comfortable and pleasant oasis is right there waiting for you. The comfortable seats by the landing window take a panoramic view while wondering off your mind to get a valuable leisure time.

Rooms and Suites

The hotel has 266 luxurious rooms and suites, each abundant in natural light and with spacious living spaces.

Ingenious in design, each room is decorated with colorful contemporary art. The headboard is meticulously hand-painted with cloud patterns, symbolizing the happy and the auspicious of traditional Chinese culture and revealing subtle and good moral. Meanwhile, the carpet in the rooms are dotted with ink cloud rolling, which with purple, ivory and creamy white, sketches out a clouding picture quietly elegant. The carefully selected furniture is luxurious, comfort, and full of texture.

DINING

Fragrant Restaurant

Just as its name implies, Fragrant Restaurant makes a fragrant collection of food, providing a good place for people who love, food and wine. Designed by world-renowned SPIN, the emphasis is laid on the interior texture and the combination of modern and traditional Chinese elements. The entrance glisters very well with light by spotlights with ceiling decorated with a wire sculpture. The sculpture has a strong artistic sense.

Excellence Pavilion

As a signature restaurant of this hotel, Excellence Pavilion takes into the account the tradition that the cuisine occasion can make a good chance for people to renew old friendships and make new ones, as well as demands for dining ritual. The art exhibitions and the overall design are unified. At the entrance of the restaurant, there are modern art pieces of colored top-level bone porcelain, which respond with the carpentry down the ceiling. Soft lighting and wood walls highlight the space privacy. Private rooms are given a warm tone, through whose landing windows come the breathtaking views of the city.

Plaza Gallery

Plaza Gallery features a stunning night view of the city, so guests get an extraordinary experience which changes between reality and fantasy. The bar makes a place of peace and quiet with its unique, exquisite and decorative design, where you are bound to be kept away from city hustling and bustling.

Banquet and Meeting Space

There are ballrooms of different styles and sizes, mainly located on the 3rd, the 7th, and the 29th floors. Those on the 7th and the 29th are called with floral names, like jasmine, magnolia, peony, and lotus. Chinese aesthetics is therefore revealed completely. On the 3rd floor, Four Seasons Ballroom has two lobbies, flexible to meet the needs for private workplace, dining room, dressing room, VIP reception or wedding and even dresser for actors. Among the conference arrangement, desk, fret, U, theater, cocktail parties, Chinese banquet and board meeting are available. With an area of 150sq.m, the Joy Terrace can be used for reception or wedding celebration, leaving guests precious memories. In addition, each ballroom of the Four Seasons Hotel in Shenzhen can meet the needs of meetings and events of different sizes. Four Seasons Ballroom can accommodate up to 360 guests, ideal for holding business meetings as well as wedding celebration. And the smallest meeting space is suitable for about 20 people to have intimate gatherings.

Sports Facilities

The hotel's 24-hour fitness center is easy and convenient for travelers to recover from jet lag, where they can take cardio and weight training, or relax in sauna, steam room or vortex pool. The indoor swimming pool is the best choice for water sports, while the larger outdoor pool is a good place both for kids and families.

融合南亚风情，书写浪漫故事
To Blend South Asian Style, to Make Romantic Story

项目名称：宝格丽巴厘岛度假酒店
设计公司：安东尼奥设计

Project Name: Bulgari (Bali) Resort Hotel
Design Company: Antonio Citterio & Partners

宝格丽度假酒店位于旅游胜地巴厘岛西南端，由 Antonio Citterio and Partners 建筑师事务所采用手工雕刻的火山石、珍贵奇异的木材以及精美的织物设计而成，既充分吸收了巴厘岛独特的建筑风格，又延续了宝格丽的意大利浪漫风情。在视觉上，整个度假酒店以取自海角的 butik 岩石制成的矮墙为特点，仿如一座隐蔽在城墙之中的中世纪城堡。

度假村由 59 栋带有嬉水池和庭院、热带风情的花园、高科技功能和各类设施的海景别墅组成，包括 3 栋双卧室别墅和 1 300 平方米的宝格丽别墅，其设计体现出现代感和国际化风格。花园和内墙采用天然熔岩和 palimanan 石装饰，而精致的 bangkiray 硬木、天然绿色的 sukabumi 石则被用于淋浴房、嬉水池和游泳池。此外，每栋别墅的卧室和浴室、餐厅和某些凉亭的门窗都以桃花心木和玻璃材质相结合，突出纯粹的西式设计，而材料则取自于当地，并在本地手工制作。在度假村的公共空间和客房内，随处可见独一无二的陶瓷以及蕴含完美工艺的实用和装饰品。这些作品来自于对材质、方法和生产工艺的准确研究，是巴厘岛为宝格丽特别设计和手工制作的。从日常生活到本地文化、宗教、艺术、手工艺和自然，印度尼西亚和巴厘岛风情为整个创作过程激发灵感，从而赋予了每件作品历史感和深刻含义。

餐饮

Il Ristorante

Il Ristorante 是宝格丽度假村独有的意大利餐厅，包含 36 个座位，以浪漫柔和的灯光营造出迷人优雅的氛围。

Sangkar 餐厅

休闲轻松的 Sangkar 餐厅位于悬崖边。Sangkar 在印度尼西亚语中表示"笼子"，餐厅以各种式样的传统鸡笼，作为照明和餐厅装饰。餐厅可容纳 70 人，室内陈列着一件 9×1.5 米令人惊叹的艺术品，由巴厘岛艺术家 Made Wianta 为 Sangkar 特别创作，采用实心柚木板和 100000 多颗铜"钉"制成。

酒吧

配有舒适躺椅的露台酒廊，是享受壮观的海洋美景与华丽日落的理想场所。酒吧的椭圆形弧形柜台令人联想起宝格丽米兰酒店的黑色树脂酒吧。在北

苏拉威西道的一次大地震后、发现的由水流冲击成的巨大天然岩石雕塑、鬼斧神工、极为震撼。

康体设施
水疗中心

水疗中心建造在狭窄葱翠山谷脚下的悬崖边缘，包括瑜伽亭、露天凉亭、漂浮式木桥，其建筑多以水池和巨大玻璃窗为特点，肌理和色彩的完美和谐，营造出富有现代感、令人放松的环境氛围。建筑将总体现代的设计风格与传统的巴厘岛元素融为一体，源自爪哇的Joglo古代建筑，也被精确的重建于此。此外、大量珍贵的当地木材，也被广泛的应用于此。从入口处的亭子通往手工雕刻的柚木Joglo建筑，它迁移自爪哇东北部的城市库达斯。外部错综复杂的雕刻反映着爪哇、中国和阿拉伯文明，还有古老的印度玛迦帕西特王朝多年来在库达斯城留下的遗迹。继续深入，内部是露天休闲室、瑜伽室和游泳池。还有淋浴房、芳香蒸汽浴室，覆盖在金色树叶状玻璃和祖母绿色彩之下。

商务与婚宴设施

拥有户外露台与观景阁、空间宽敞的多功能厅以及优雅的会议室构成了度假村商务与婚宴设施。多功能厅直接通往具有壮丽海景的宽阔露台区域，最多能容纳100人同时用餐，专为社交聚会与会务而设计。
宝格丽别墅中底层花园设有凉亭，而上层草坪则设计有巨大的嬉水池，是户外婚礼宴客的完美之所，最多可容纳50人。

Located in the southwest tip of Bali Bulgari Resort Hotel, is endowed with hand-carved volcanic stone, precious exotic wood and fine fabrics by Antonio Citterio and Partners Architects, where to fully absorb the unique architecture of Bali style in continuing the romantic Italian style of Bulgari.

Visually, the entire resort features sunk fence of Butik stone taken from the cape of Bali Island, like a medieval castle hidden in walls. The resort constitutes of 59 ocean view villas, where to accommodate wading pool, courtyard, tropical garden, high-tech features and various facilities. Additionally, there are three double-bedroom villas and Bulgari villa of 1,300 square meters. All reflect a modern and international style. The garden and the interior wall are decorated with natural lava stone and palimanan, while delicate bangkiray hardwood and sukabumi are used to wrap shower room, splash pool and pool. Furthermore, bedrooms, bathrooms, dining rooms and doors and windows of pavilions, have adopted mahogany and glass to highlight a purely western design. All materials come from and are handmade in the local place.

In public spaces and guest rooms, everywhere are unique ceramics and practical decorations. These works are accomplished with accurate study of materials, methods and manufacturing processes. From everyday life to local culture, religion, arts, crafts and nature, themes of Indonesia and Bali offer inspiration for the whole process, providing a sense of history and deep meaning for all pieces.

DINING

Il Ristorante

Il Ristorante is a unique Italian restaurant, with 36 seats in a charming and elegant atmosphere of romantic soft lighting.

Sangkar Restaurant

Located on the edge, Sangkar restaurant is casual and relaxing. In Indonesian language, Sangkar means cage. And in this dining space, a variety of traditional cages now serve as lighting and decoration. The space accommodating up to 70 people has an art piece measuring 9 meters long and 1.5 meters wide. An item it is with solid teak panels and more than 100,000 copper nails, specially made by local artist Made Wianta.

Bar

Bar enjoys spectacular scenery and has comfortable lounge chairs on the terrace lounge, making an ideal place to appreciate the spectacular ocean views and gorgeous sunsets. The oval curved bar counter is reminiscent of the black resin bar in Bulgari Hotel, Milan. A natural huge rock here becomes a major sculpture, done with water hitting and found after a major earthquake in North Sulawesi Road.

SPORTS FACILITIES

Spa

At the foot of lush cliff, Spa built in a narrow valley, has a yoga pavilion, an open-air pavilion, and a floating wooden bridge. An enjoyment it is featuring pool and huge glass windows everywhere. A variety of surfaces and colors in perfect harmony brings forward a very contemporary and relaxing ambience. Here reflects an overall style that blends modern design with traditional Balinese elements. Ancient Joglo building Javanese has now been restored. Additionally, precious local wood is widely used throughout the resort.

The entrance pavilion leads to the Joglo building with hand-carved teak, a building from a northeastern Javanese city. Its external intricate carvings reflect civilizations of Java, China and Arab. Remains by Maacah Pacific Dynasty are also available here. Inside there is an open-air lounge, a yoga room and a swimming pool. Shower, and aroma steam bath are roofed with covered with golden leaf-shaped glass and coated in emerald color.

Business and Wedding Facilities

Outdoor terrace with viewing pavilion, spacious function rooms and an elegant conference room make the business and wedding facilities. The multi-purpose hall takes direct access to the sea-view terrace the dining room accommodating up to 100 people at dinner. This is specially designed for social gatherings and conference.

樱花绽放迎嘉宾
Cherry blossom to Greet Guests

项目名称：东京半岛酒店

Project Name: The Peninsula, Tokyo

东京半岛酒店坐落于东京丸之内商业区，与京都御花园隔街相望，距离银座购物中心仅有数分钟路程。它以优越的城市景观、奢华舒适的住宿、精良先进的设施、品质非凡的餐饮和富有传奇色彩的半岛酒店服务而著称。

酒店拥有全城无可比拟的美景，皇宫花园和日比谷公园就近在眼前。知名室内设计师桥本夕纪夫选用大地色、木材、漆器和大理石进行混搭，并在设计中充分考虑到各项设施的功能性，为宾客营造奢华舒适的居住环境，被赞为"国际设计闪耀日式精彩"。

东京半岛酒店汇萃传统及新派美食精华，著名的大堂茶座下午茶、起凤台高级粤菜、京都 Tsuruya 日本料理，各具风味，提升美食境界。顶层 Peter 餐厅呈现皇宫御花园和日比谷公园优美景观，是商务午餐或银座购物之后稍作休息的绝佳场所。

在日本，樱花季是全年景色最迷人的时刻，每年都能吸引数百万游客前来参观，而且樱花也是无数艺术家、摄影师、电影制作人和诗人为之沉醉的主题。2013年，东京半岛酒店悉心准备精美装饰，精心策划主题优惠活动，力邀八方宾客，共同体验前所未有的灿烂樱花季。3月8日至15日，东京半岛酒店在大堂茶座中摆放了各式迷人的樱花装饰，与室外怒放的美丽樱花交相呼应。其中包括七组高度超过2.5米的樱花立饰、64盆樱花桌面摆花以及10盏专门定制的日式灯笼。大堂茶座标志性吊灯烟花由日本艺术家滨惠泉设计而成，所有这些樱花装饰在美丽吊灯照射下更显夺目光彩。

Opposite the Kyoto Imperial Garden, "The Peninsula Tokyo" is located in Marunouchi business district with just a few minutes' distance away from the Ginza shopping district. Tt has always been well-know for its great city view, luxurious comforts, sophisticated facilities, extraordinary dining and good service.

The hotel has an unparalleled view with Imperial Palace Garden and Hibiya Park in sight. Yukio Hashimoto, a famous interior designer, employed earth colors, wood, lacquer and marble into functional facilities to create a luxurious and comfortable living environment, which won it an honor as "International Design and Japanese Marvel".

Here offers cuisines of traditional and new style. With unique flavor, the afternoon tea in the lobby, the senior Cantonese food, and the Kyoto Tsuruya further enhance the cuisine to a higher level. When introducing scenes from Imperial Garden and Hibiya Park, the Peter on the top is a perfect place for a business lunch or a break after shopping in the Ginza.

In Japan, the season when cherry is blossom is the most fascinating moment year-round, attracting not only millions of tourists to visit, but also numerous artists , photographers, filmmakers and poets. In 2013, The Peninsula Tokyo was carefully prepared, after being beautifully decorated and well-planned, to invite guests to experience an unprecedented brilliant cherry blossom time. From March 8 to 15, a variety of stunning cherry decors were placed in the lobby to echo with the beautiful cherry blossoms outdoors, like seven cherry groups over 2.5 meters higher, 64 pots placed on table and 10 customized Japanese lanterns. The iconic chandelier is designed by a famous Japanese artist. With light cast on, all are more dazzling in brilliance.

Feel the Eastern Zen Style — Living Space X · 299 ·

西子湖畔的繁华隐贵
Reclusive Noble on the Bank of Xihu Lake

项目名称：杭州逸酒店
设计师：钟刚勇

Project Name: The East Hotel in Hangzhou
Designer: Zhong Gangyong

杭州逸酒店是一家品位独特的精品酒店，是杭州新贵精品酒店标杆之作，南接武林、北瞰湖墅、东连运河与钱江新城、西望黄龙与西溪湿地，迷人西湖咫尺可达。

逸酒店的设计灵感来自于杭州四季的逸致情趣，把杭城的闲适、优雅、逸趣融入了现代的设计当中，以宁静舒适的视觉感受和低调的奢华的精神内涵，满足了人们对品位生活的向往及宁静舒适"家"的感受。从唯美的水景门厅到古朴的漆艺前台、奢华的公共摆设；从动态的"富春山居图"到电梯层的日月盈亏、电梯轿厢内的3D森林；从精致温馨的壁炉到三潭映月的吊灯、弥足珍贵的中外典籍；从手工刺绣的杭州丝绸屏风到文房四宝、墙上的古琴；从葱郁的竹林到潺潺流水、露天藤椅、每一处都体现了"大逸逸于世"的概念，为宾客营造舒适惬意的休闲空间和无与伦比的奢华享受。

漫步进入房间，您会在不经意间发现门口房号上面的新月、墙面的杭州丝绸配手工刺绣、触手可及的书籍装饰画、电视机背后的西湖全景图以及窗前市中心繁华城市景观。设计师以杭州的四季颜色为基础，选取了杭州四季中最具代表的四种植物春日龙井、夏日睡莲、秋日桂花、冬日梅花融入到酒店的设计中，通过视觉感官，让宾客对酒店有一个清晰的概念。

逸酒店超越了传统意义上的奢华酒店，它打破成规、倡导舒展松弛、拒绝乏味单调、臻于细节雕琢、让人置身于逸酒店有如置身于家中一般轻松自在，造就出非凡的体验。

The East Hotel in Hangzhou is a unique boutique hotel, a benchmark for upstarts. It enjoys a good traffic convenience with Wulin on south, Hushu on north, the Grand Canal on east and the Qianjiang River New City, and Huanglong and Xixi Wetland on west. Additionally, the Xihu Lake is right within its reaches.

Its design inspiration comes from the carefree mood of the local seasons. The quiet and elegant ease is blended the modern design. The comfortable visual experience and the understated luxury, meet people's yearning and desiring for a taste of life and a quiet and comfortable home. All sections embody a concept that urban ease can make great ease, from the water gate in the foyer to the beautiful ancient lacquer reception desk and the lavish public display, and from the exquisite cozy fireplace to the chandelier of Three Pools Mirroring the Moon. All create a safe and comfortable space and offer an unparalleled luxury Moon Lake, including precious and foreign classics, hand-embroidered silk screens of the local four treasures, lush bamboo forest, gurgling water, and outdoor wicker chairs.

Walking into the room, you will run into a crescent above the room number, you will find the door of room number above, silk of hand embroidery on the wall, decorative painting books, the Lake Panorama behind TV and the downtown cityscape by the window. Seasonal colors of Hangzhou are implanted within, like Longjing, a famous green tea in spring, water lily in summer, fragrans in autumn and winter plum in winter. Such visual senses, allows guests to have a clear concept of the concept.

A hotel this project is that has gone beyond the traditional sense of a luxury hotel. It breaks away conventions, and promotes relaxation. It denies boring tedium but lays significance on details. People therefore feet at home, getting a remarkable experience once here.

热情西藏风情，跃动多姿文化艺术

Tibet of Ardor Motivates Diversity of Culture and Art

项目名称：拉萨香格里拉大酒店

Project Name: Shangri-La (Lhasa) Hotel

2014年，雄踞于世界之巅的拉萨香格里拉大酒店隆重开幕。酒店匠心独运，深得西藏文化精髓，以热情奔放的民风、跃动多姿的文化艺术为核心设计元素，通过结合藏族风情和香格里拉酒店的现代豪华设施，完美地诠释藏式建筑。内部空间采用当地艺术品、手工艺品和纺织品进行装点，以适当地平衡现代与传统元素。香格里拉品牌源自詹姆斯·希尔顿（James Hilton）著于1933年的小说——《消失的地平线》中所描述的一处传奇圣地。

酒店门廊采用气派典雅的藏式大门设计，门身为当地工匠手工雕琢及涂绘而成。门廊的规模在拉萨地区可谓首屈一指，象征着宁静和平和，彰显了香格里拉文化与西藏精神的完美融合。墙体外观效仿西藏传统建筑，而朴实无华的朱红拱门、精雕细琢的青铜门环与门板相得益彰。门框刻有传统藏式建筑常见的精致饰纹。踏入门廊，丰富多样的本地植物和花卉即映入宾客眼帘，并一直延伸至酒店主门。酒店大门面朝布达拉宫，而牦牛油灯及亲水元素的运用为酒店增色不少，也与当地寺庙建筑有着异曲同工之妙。手绘门廊以及地板上均刻有藏式图案以及两种寓意吉祥的藏族象征——莲花和吉祥结。

穿过酒店一层即可通往一座天然绿洲般的酒店花园。园内种满各种当地植物，群芳争艳，绿树环绕，空气格外清新。花园深受传统藏式"林卡"（Linka）园林设计的启发，园内随处可见色彩艳丽、象征着幸福的西藏格桑花。除小池塘和喷泉之外，园内还设有优雅宜人的藏式亭台。

客房及套房

酒店内设262间现代客房及17间宽敞套房。客房室内环境舒适，面积42（43）平方米起。套房空间奢华大气，面积86平方米起。宾客可一边欣赏醉人的布达拉宫景致、令人叹为观止的连绵群山或花园美景，一边享受设计现代、荟萃各种藏族元素的舒适客房。所有双床客房均采用生机勃勃的绿色及蓝色，而特大床客房则采用质朴的橙色和前卫的蓝色，套房则以淳朴的棕色和米白色为主

色。所有客房均采用吉祥结、卷云、花卉等藏式图案装点。客房装饰及枕套的风格深受藏族头饰及宝石的影响。

所有客房及套房均提供免费 WiFi 网络及 40 英寸卫星有线液晶电视、迷你酒吧、长沙发和酒吧桌等设施。现代浴室采用 Toto 浴具，淋浴间则采用吉祥结装饰，并配有一个藏式洗手盆。所有套房及公寓均配有浴缸、浴室电视、迷你酒吧及橱柜。

餐饮

驿站

驿站是大堂的焦点所在。酒廊位于酒店第三层，部分墙体采用透明玻璃幕墙设计。酒廊格局取材于最能体现西藏精神的曼陀罗图案。室内的温暖色调以及质朴色泽深受传统唐卡绘画的启发。

位于驿站酒廊中央的吊灯堪称点睛之笔。吊灯长 6.5 米，气派非凡，让宾客不

禁驻足仰望。吊灯底下设有一座高台，而钟形吊灯则仿如一座转经轮。吊灯还采用了大量的朱红布料及青铜配饰，形似西藏经幡。另有一面巨大的"8"字形金色网格徽章高挂在墙上。中国文化中，数字"8"寓意着幸运和财富。宾客可在酒廊高挑的天花板下休息，一边感受微风吹拂，一边品尝各种小吃和饮品；亦可在户外庭院放松身心，远眺连绵群山。

酒店第三层的接待厅背后的浮云装饰为酒店空间营造出梦境般效果。卷云是一种古老的藏族符号，寓意团结和力量，在这里则化身成一件由青铜锤铸而成当代艺术作品。卷云层次丰富，质感细腻，与光滑的深棕色墙体形成了鲜明的对比。

云顶

云顶是酒店开设的全天候餐厅，与驿站共享一个庭院。宾客可在此品尝丰富的自助早餐或晚餐；午餐采用单点形式，为宾客提供品种丰富的美食。餐厅设计现代，藏式木刻以及箱架的运用更是锦上添花。

云顶餐厅供应各种当地特色菜肴和国际美食。厨房采用开放式布局，是香格里拉酒店集团享誉世界的特色概念。冷盘及海鲜档供应各种新鲜进口海产、进口奶酪、风味独特的本地奶酪和现制冷盘。烧烤档供应现烤牦牛排、日式烧鸡、烤海鲜及其他烧烤特色美食。

香巴拉

香巴拉位于酒店三层，这间别致的酒廊式餐厅内设多种坐席选择，分为餐饮区、休息区、酒吧区及户外庭院区。店内滇藏风情美食云集，特色名菜有红烧藏式奶酪腌羊羔肉以及熏肉酿米丸子。

宾客还可尽情品尝丰富多样的传统云南美食。餐厅采用暖色内饰，灯饰仿佛僧人手持的油灯，散发出朦胧的灯光，而墙上挂着各式藏族手织毛毯。其他藏族元素还有低天花梁金属装饰以及随处可见的吉祥结。

香宫

香格里拉特色餐厅——香宫以正宗的川粤美食以及现代时尚的进餐环境为宾客呈献独特的珍馐体验。餐厅中的传统中式和藏式风情水乳交融，营造出一种宾至如归氛围，让宾客细尝各种佳肴。另设10间内饰豪华的包厢，适合私人聚会。

康体设施

氧吧

拉萨雄踞在青藏高原上的萨河北岸，海拔3 650米，在藏语中意为"佛地"，是西藏的首府和规模最大的城市，以及地球上海拔最高的人类居住地之一。为让客人适应拉萨的高海拔环境，拉萨香格里拉大酒店将设有氧吧，开创当地五星级国际酒店先河。

氧吧内的氧气浓度与海平面相当，周种植了丰富的当地植被以提高含氧量，宾客可在氧吧中恢复体力、享受足底按摩及美甲服务。

"气"Spa

酒店设有香格里拉的特色"气"Spa，宾客可在此尽情放松、享受特色按摩、面部美容和多种养生疗法。这里是宾客探秘拉萨后恢复充沛精力的绝佳之所。水疗馆内设10间宽敞舒适的私人理疗室，每间理疗室均以十大藏族吉祥物之一命名。

健身中心

对于已适应高原环境并想健身锻炼的宾客，酒店开设了一间设备齐全、带22米室内泳池的健身中心。室内采用大幅玻璃窗设计，宾客可在尽情畅泳或在池畔休闲放松之余欣赏花园景致。

宴会及会议场地

酒店内设宴会和活动空间，是宾客举办大型会议及活动的不二之选。宴会厅楼高两层，是全拉萨城内规模最大的五星级国际会议及宴会设施。不论宾客的要求如何，酒店的专业活动管理团队都将全力协助，确保活动完满举行。

大宴会厅

大宴会厅面积1 215平方米，层高6.8米，规模属全城最大。踏入迎宾门廊，宾客可观赏到手工雕琢的天花板装饰、吊灯及精致配饰，以及可远眺布达拉宫的小型露台。

从气派堂皇的碧绿色大门进入大宴会厅后，一幅令人叹为观止的艺术作品映入眼帘——宴会厅内的平面莲花彩绘。画作上的蓝色和金色与绣有吉祥结图案的毛毯相得益彰。大宴会厅的格局深受大昭寺启发，以涂金墙纸覆盖天花板。从大宴会厅的露台也可眺望远山，此处风景也属全城独有。

宴会及多功能厅

所有宴会功能厅均以西藏各大山脉命名。位于酒店五楼的休息区通往露台，饱览布达拉宫以及四周延绵山峦风光，是举行宴前鸡尾酒会的绝佳之所。

商务中心

位于酒店第三层中心位置的商务中心设备齐全，且为宾客提供便捷的旅游交通信息等一站式自助服务。商务中心设计现代，功能齐备，设有办公室及免费宽带网络连接，能提供各种商务支持服务。

Resting on the highest peaks in the world and opened in 2014. Shangri-La Hotel, Lhasa, the Shangri-La Hotel, Lhasa, was carefully planned with a deep appreciation for Tibetan culture. The vibrancy of the people, culture and art is the cornerstone of the hotel's design and concept. The hotel's design is a conscious blend of Tibetan heritage and modern Shangri-La luxury. The Shangri-La brand was inspired by the legendary land featured in James Hilton's 1933 novel, Lost Horizon.

The hotel's entrance features a grand Tibetan gate that was hand carved and painted locally. The gate, one of the largest in Lhasa, symbolizes tranquility and serenity. It shows a magnificent blend of the Shangri-La culture and Tibetan spirit. The walls are plastered in traditional Tibetan style, while earthy red arches complement the delicately carved bronze doorknockers and panels. The frame is carved with intricate details of traditional designs commonly found in Tibetan architecture. Upon entering the gate, guests will find a variety of indigenous plants, flowers and greenery as they drive up to the main entrance.

The main entrance of the hotel faces Potala Palace and is enhanced with burning yak butter lamps and water features. These are similar to the features found at local temples and monasteries. The hand-painted porch is carved with Tibetan motifs and two of the most auspicious Tibetan symbols, the lotus and the eternity knot, are etched on the ground.

Accessible via Level 1, the garden is a natural oasis filled with indigenous plants, flowers and trees that create

oxygen. The garden is inspired by traditional Tibetan gardens called Linka. Throughout the garden, guests will find the Tibet's Gesang flower. These brightly colored flowers symbolize happiness and good luck. Alongside small ponds and fountains, Tibetan pavilions grace the relaxing environment.

A floating cloud installation behind the Level 3 reception desk transforms the space into a dream-like state. Signifying unity and strength, the ancient Tibetan symbol of swirling clouds is made into a contemporary work of art using hammered oxidized bronze. The layered clouds with vast detail contrast the smooth texture of the dark brown walls.

GUST ROOMS AND SUITES

The hotel comprises 262 contemporary guestrooms and 17 expansive suites. Rooms are cozy with a minimum size of 43 square meters, and suites starting from 86 square meters offer the luxury of space. Guests can enjoy stunning views of Potala Palace, and the breathtaking mountains or the garden in the comfort of their modern-design room infused with Tibetan elements. All twin-bedded rooms are accented with vibrant greens and blues, while king-bedded rooms are accented with earthy oranges and bold blues. The suites are toned down with earthy browns and beiges. All rooms are decorated with traditional Tibetan motifs, including endless knots, swirling clouds and floral patterns. The room fixtures and pillows are inspired by Tibetan headdress and precious stones.

All guestrooms and suites include complimentary Wi-Fi Internet access, satellite cable on 40-inch LED televisions, mini bar, sofa settees, lounge tables and much more. The modern bathrooms are fitted with Toto washlets,

while the shower cubicle is decorated with an endless knot motif. Every shower cubicle is fitted with a washbasin that caters to guests' needs as they acclimate to the altitude. All suites and apartments are fitted with bathtubs and televisions in the bathroom, as well as a decorative mini bar and cabinet.

DINING

Lodgers Lounge

Lodgers Lounge is the epicenter of the hotel lobby. Located on Level 3, the lounge is partially surrounded by sheer screens. The concept for the floor plan derives from one of the most spiritual references in Tibet, the Mandala. The warm tones and earthy colors are inspired from Thangka, a traditional painting.

The centerpiece of Lodgers Lounge drawing guests to the center of the room is an astonishing 6.5-metre chandelier. With the raised platform below, the bell-shaped chandelier is reminiscent of a prayer wheel. It is engulfed in red fabric with bronze details, symbolizing Tibetan prayer flags. The large gold mesh medallions on the wall are in the shape of the number eight, which signifies fortune and wealth in Chinese culture. Guest can relax under high ceilings and enjoy the breeze as they savor snacks and beverages in the lounge. They can also unwind on the outdoor patio, while taking in the mountain views.

Altitude

Sharing a patio with Lodgers Lounge is Altitude, the hotel's all-day dining restaurant. Diners can enjoy a lavish buffet for breakfast and dinner. Lunch will be strictly a la carte with a variety of options. Hints of Tibetan wood carvings and chest-like shelving highlight this contemporary-design restaurant.

At Altitude, guests will find a variety of local and international cuisines. The live-action, open-show kitchen format is a signature concept of Shangri-La Hotels and Resorts worldwide. The Cold and Seafood Station features fresh imported seafood, imported cheese, special local cheese and cold cuts sliced on the spot. The Grill Station heats things up with yak steaks, yakitori, grilled seafood and other grill specialties.

Shambala

Also located on Level 3 is the hotel's signature dining concept, Shambala. The bar and lounge features many seating options; a dining area, lounge, bar seats and outdoor patio. The Tibetan-Yunnan tapas bar and lounge's menu pay homage to the region by featuring authentic Tibetan cuisine. Signature dishes include Roast Lamb marinated in Tibetan Yoghurt and Baked Bacon wrapped in Ground Rice Cubes.

Guests will also find a variety of traditional Yunnan dishes on the menu. The restaurant's warm interior is dimly lit by lights reminiscent of oil lanterns carried by monks, while the walls are draped in Tibetan hand-crafted carpets. Other Tibetan influences include decorative metal fixtures across low ceiling beams and the repetitive use of the endless knot.

Shang Palace

Shang Palace, a Shangri-La signature restaurant, serves authentic Cantonese and Sichuan cuisines in a contemporary setting. The restaurant blends traditional Chinese and Tibetan styles and provides guests an inviting space to enjoy a meal. Ideal for intimate gatherings, the restaurant offers ten lavishly decorated private dining rooms.

The Oxygen Lounge

Lhasa is the capital and largest city in Tibet and means "Place of God". It sits on the Qinghai Tibetan plateau on the banks of the Lhasa River at an altitude of 3,650 meters above sea level, and is one of the highest inhabited regions on earth. With high altitudes in mind Shangri-La Hotel, Lhasa is the first five-star international hotel to introduce an Oxygen Lounge.

The concentration of oxygen in the lounge is equivalent to sea level. The lounge is surrounded by extensive garden landscaping featuring indigenous plants to increase oxygen levels. Inside the oxygen lounge, guests can recharge and enjoy reflexology treatments, manicures and pedicures.

CHI, The Spa

The hotel has a Shangri-La signature outlet, CHI, The Spa. At CHI, guests can unwind with one of the spa's signature massages, facials or treatments.

An ideal space to rejuvenate after a long day of exploring the city, the spa offers guests comfort in one of ten private rooms. Keeping with the original CHI Spa concept, the private rooms are all named after ten Tibetan auspicious symbols.

Health Club

For guests who have acclimated and want to hit the gym, the hotel has a fully equipped gym with a 22-metre indoor swimming pool. From large glass windows, guests can enjoy the garden view as they swim or relax by the pool.

BANQUETING AND CONFERENCE VENUES

For destination meetings and events, organizers will find ballrooms and function spaces ranging from 81 to 1,215 square meters. Spread across two floors, the hotel offers the city's largest five-star international conference and banqueting facilities. Whatever the requirement, the hotel's professional Events Management team will always be on hand to ensure the successful conclusion of events.

Grand Ballroom

Spanning over 1,215 square meters with 6.8-meter ceilings, the Grand Ballroom is the largest in the city. Upon entering the foyer, guests will find hand-carved ceiling decorations, chandeliers with exquisite details and a small terrace with Potala Palace views.

Guests will then enter through the grand turquoise-colored doors and be greeted by breathtaking artwork. The two-dimensional lotus painting immediately draws them into the ballroom. The blue and the gold of the painting complement the carpet that carries the endless knot motif. The layout of the Grand Ballroom is inspired by the Jokhang Temple, while the ceiling is covered in wallpaper using a gold brush technique. The Grand Ballroom is also the only Grand Ballroom in the city with a quaint terrace that offers mountain views.

Function Rooms

All the function rooms are named after mountains in the region. The breakout area on Level 5 leads to the outdoor terrace, which overlooks Potala Palace and the surrounding mountain range and makes it an ideal spot for pre-event cocktails.

Business Center

Centrally located on Level 3, it is a one-stop self-service lounge providing guests comprehensive and convenient information on tours and transportation. It also offers business support services. This modern space is fully equipped with work stations and complimentary broadband Internet access.

图书在版编目（CIP）数据

禅意东方 X 居住空间 / 黄滢，马勇 主编 . – 武汉 : 华中科技大学出版社，2014.7
ISBN 978-7-5680-0279-0

Ⅰ . ①禅… Ⅱ . ①黄… ②马… Ⅲ . ①住宅 – 室内装饰设计 – 作品集 – 世界 Ⅳ . ① TU241

中国版本图书馆 CIP 数据核字（2014）第 170921 号

禅意东方 X 居住空间

黄滢 马勇 主编

出版发行：华中科技大学出版社（中国·武汉）	
地　　址：武汉市武昌珞喻路1037号（邮编：430074）	
出 版 人：阮海洪	
责任编辑：熊纯	责任监印：张贵君
责任校对：段园园	装帧设计：筑美空间
印　　刷：中华商务联合印刷（广东）有限公司	
开　　本：965 mm × 1270 mm　1/16	
印　　张：20	
字　　数：160千字	
版　　次：2014年10月第1版 第1次印刷	
定　　价：338.00元（USD 67.99）	

投稿热线：（020）36218949　　duanyy@hustp.com
本书若有印装质量问题，请向出版社营销中心调换
全国免费服务热线：400-6679-118 竭诚为您服务
版权所有　侵权必究